W9-AEU-951

WHERE DO COMETS COME FROM?

HOW HAVE THEY AFFECTED EARTH?

AND WHAT IS THERE TO HOPE AND TO FEAR FROM THEM?

Curiosity about comets has always burned as bright as their fiery glow emerging from the darkness of space.

What we have learned and still are learning about them is more intriguing still.

This absorbing account answers many of our questions about what makes these celestial eccentrics tick, and why their strange behavior has fascinated astronomers for thousands of years.

COMETS
———THE
——SWORDS
——OF HEAVEN

DAVID RITCHIE is a highly acclaimed professional science writer. His work has appeared in numerous publications, among them *Analog, The Boston Globe, Newsday, The New York Times Magazine,* and *The Washington Post.* He is the author of *The Ring of Fire* (available in a Mentor edition), *Spacewar* (available in a Plume edition),and *The Binary Brain.*

COMETS
THE
SWORDS
OF HEAVEN

David Ritchie

A PLUME BOOK

NEW AMERICAN LIBRARY

NEW YORK AND SCARBOROUGH, ONTARIO

PLUME TRADEMARK REG. U.S. PAT. OFF. AND FOREIGN COUNTRIES
REGISTERED TRADEMARK—MARCA REGISTRADA HECHO EN
WESTFORD, MASS., U.S.A.

SIGNET, SIGNET CLASSIC, MENTOR, PLUME, MERIDIAN
AND NAL BOOKS are published *in the United States*
by New American Library, 1633 Broadway, New York, New York 10019,
in Canada by The New American Library of Canada Limited,
81 Mack Avenue, Scarborough, Ontario M1L 1M8

Library of Congress Cataloging in Publication Data

Ritchie, David, 1952 Sept. 18-
    Comets, the swords of heaven.

    Bibliography: p.
    Includes index.
    1. Comets. 2. Halley's comet. I. Title.
QB721.R64 1985   523.6   84-27284
ISBN 0-452-25619-4

First Printing, June, 1985

1 2 3 4 5 6 7 8 9

PRINTED IN THE UNITED STATES OF AMERICA

To Carol Mann

Preface

The trouble with many books about comets is that they fall into one of three categories. The first is the dull, comprehensive text full of dry facts and figures; the second appears to have been written for a readership of second-graders and is packed with cute silliness; and the third is the coffee table book full of pretty pictures with an unenlightening text. With the return of Halley's Comet upcoming, we may expect to see additions to each of these categories. This book is an effort to avoid their failings by going beyond the boring lists of data and the gorgeous photos and showing the reader how important a place comets occupy in human history and specifically in the history of science. If comets had never existed, the world we live in might be quite a different place, for comets have served as triggers for wars, mass extinctions, and quantum advances in human knowledge; indeed, no other celestial bodies, with the exceptions of the sun and moon, have done more than comets to make us what we are and our civilization what it is. That statement may strike you as excessively broad and sweeping now, but your opinion will probably change by the time you finish the final chapter. One of the aims of this book is to demonstrate how everything in our corner of the universe, from the reigns of kings to the tails of comets to the particle output of the sun, is connected, directly or indirectly, with everything else; and I hope readers will see how humankind and comets serve as pieces of the vast jigsaw puzzle that is our universe.

Not everything here will necessarily interest everyone. Chapter 5 treats the physics of comets in detail, and some readers may wish to ignore it. For such readers, the rest of the book is written so that an understanding of Chapter 5 is not essential. So you may skip it if you like and rely on the Glossary for explanations of unfamiliar terms. You may want at least to sample that chapter, however, for it illustrates how the scientific method has been applied to comets—not always successfully—and how scientists looking at the same body of data can come up with dramatically different conclusions.

A book that dealt with comets to the exclusion of all else would be both short and extremely dull and would not provide much of an education for the reader. So I ask the reader's patience with what may seem occasional short digressions from the narrow topic of comets, for the side trips will not last long, and they will offer fresh ways of looking at this familiar phenomenon in the skies.

Contents

One: Comet Lore and History 1

Two: The Comet Hall of Fame 32

Three: The Rise of Comet Science 44

Four: Newton, Halley, and the Comet of 1680 67

Five: Comet Science Comes of Age 84

Six: A Comet Ride 114

Seven: The Siberian Comet Mystery 130

Eight: Comet Impacts 143

Nine: Comets and Catastrophes 155

Ten: Myths, Disease, and Comet Dust 171

Eleven: How to Watch for Halley's Comet 199

Twelve: Scientific Studies of Halley's Comet 209

Glossary 214

Who's Who in Comet Science 221

Recommended Readings 225

Index 228

COMETS

THE
SWORDS
OF HEAVEN

ONE

Comet Lore
and History

COMETS AND
"THE UNIVERSALITY OF THINGS"

"It appeared to be of great length [and] the color of blood. At its summit was visible the figure of a bent arm, holding in its hand a great sword as if ready to strike. . . ." Thus the famous French surgeon Ambroise Paré described a comet that appeared over Europe in 1528. Paré was only a boy when he witnessed the comet, but the sight impressed him so deeply that many years later he remembered it in detail. Paré described the comet's "point," or head, as having "three stars," while on either side of the tail "were seen a great number of axes, knives, and bloodstained swords, among which were . . . hideous human faces with beards and bristling hair." Though most of these features existed only in Paré's imagination, his reaction to the comet was mild compared to the behavior of many other people. Some individuals were literally scared to death. "The comet was so horrible . . . and produced such great terror among the common people," Paré observed, "that many died of fear and many others fell sick."

Accounts like this are nothing unusual in the history of comet studies. For thousands of years, comets have been associated with all manner of disasters and misfortune. This association is easy to understand. Comets are frequent visitors to our skies, and the brightest of them are so striking that astronomer Sir John Herschel wrote in his *Treatise on Astronomy* in 1836:

> That feelings of awe and astonishment should be excited by the sudden and unexpected appearance of a great comet, is no way surprising; being, according to the accounts we have of such events, one of the most brilliant and imposing of all natural phenomena.

Small wonder, then, that when visits of spectacular comets coincided with some catastrophic events in history, a link between comets and calamities was formed in human minds. Comets have been blamed for plagues, earthquakes, floods, all manner of other natural disasters, and so many wars that they were once dubbed the "swords of heaven." Tradition has held comets responsible for death and destruction, disease and decay, defeat and dissolution, the demise of kings, and the fall of mighty empires. No celestial phenomenon has been more widely feared or more intensively examined, and the study of comets has helped to dispel many of the superstitions that have surrounded them, while advancing the science of astronomy and helping to divorce it from the ancient pseudoscience of astrology. Comets flit like swallows through the history of science, and their influence is not confined to astronomy; scientists in many other fields of study have recently come to realize how truly the English philosopher Roger Bacon spoke when he said, seven centuries ago, that "comets have some action and effect on the universality of things." Knowledge about comets has given us a better understanding of other sciences from biology to geology and has provided an explanation for many mysteries of nature here on earth. Thanks to comet studies, we now have a clearer understanding of how our own world has evolved and what may happen to it in the future. On the nonscientific side, comets have enriched our language and literature and have added a whole extra dimension—sometimes tragic, sometimes comic—to human history. Comets have engaged the attention of many of the most prominent men and women in every century, who have made the history of "cometology" a tale of saints and scoundrels, geniuses and crackpots, and a steady (though not always successful) effort to make hypothesis and theory fit the facts. That story is the subject of this book; and since comet science arose from superstition, let us start by looking at some of the strange beliefs associated with comets.

COMETS AS OMENS

Western literature is full of ominous references to the woes associated with comets. John Milton compared the devil to a comet in *Paradise Lost:*

On the other side,
Incensed with indignation, Satan stood
Unterrified, and like a comet burned
That fires the length of Ophiuchus huge
In the arctic sky, and from his horrid hair
Shakes pestilence and war.

A more gentle reminder of the reputation of comets may be seen in poet Andrew Marvell's description of glowworms:

Ye country comets that portend
No war nor prince's funeral . . .

Shakespeare, in his tragedy *Julius Caesar,* warns of comets that portend the deaths of rulers:

When beggars die, there are no comets seen;
The heavens themselves blaze forth the death of princes.

Pulpit orators have sounded anticomet warnings even more often and more loudly than poets have. The Venerable Bede warned his audiences in seventh-century England that "comets portend revolutions of kingdoms, pestilence, war, winds, or heat." Martin Luther once preached in a sermon: "The heathen say that comets arise from natural causes, but God creates not a single [comet] that does not foreshadow a certain calamity." Luther's less flamboyant but equally distinguished friend, theologian Philip Melanchthon, also believed that God sent comets to serve as scourges on a sinful world, and the views of Luther and Melanchthon inspired two Protestant preachers in Switzerland to write a rhyme about comets for schoolchildren, here freely translated from the original German:

[These] things a comet brings . . .
Storm, plague, famine, death of kings,
War, earthquake, flood, and upheaval.

On at least one occasion in the fifteenth century, when a huge comet had much of Catholic Europe terrified, the text of the *Ave Maria* was rewritten to include a prayer for protection from comets ("From the Devil . . . and the comet, Lord, deliver us"), and one ecclesiastic explained the reason why: "Comets signify corruption of the air. They are signs of earthquakes, of wars, of changing of kingdoms, great dearth of food; yea, a common death of man and beast from pestilence." For centuries afterward, the small cakes sold at the front doors of cathedrals in Italy were called *cometes* in memory of the terror that the comet caused.

Colonial America's religious leaders generally shared the traditional fear of comets. The appearance of a very bright comet in 1682 inspired Massachusetts Puritan leader Increase Mather to preach a sermon with the impressive title of "Heaven's Wrath—Alarm to the World—wherein it is shown that fearful sights and signs in the heavens are the presages of great calamities at hand." Mather thundered:

For the Lord hath fired his beacon in the heavens among the stars of God there. The fearful sign is not yet out of sight. . . . See the sword blazing over us? O pray unto Him, that He would not take away stars and send comets to succeed them!

On the secular level, Jonathan Swift described the prophetic significance of comets when he wrote that "old men and comets have been reverenced for the same reasons: their long beards, and their pretences to foretell future events." Historian Nicephoras Gregoras mentioned how "a great danger to the people [could be] predicted" from comets; and Nicephoras merely reported a practice that had been popular since the eleventh century B.C., when Chinese astronomers viewed comets as omens and tried to divine the future from their appearances. True to traditional Chinese thinking—the written characters for "disaster" mean both "danger" and "opportunity"—the Chinese thought that comets forewarned of events that might mean woe for some but a lucky break for others. Chinese astronomers tried to figure out what unlucky souls might stand on the receiving end of this celestial finger of fate, because the information might prove useful to the emperor in planning military campaigns.

MOCTEZUMA'S COMET

Comets also played a small but significant role in the downfall of the empire of the Aztecs in Mexico. About the time Hernando Cortés was putting together an expedition to conquer the Aztec realm for Spain, several comets appeared in the Mexican skies and caused the Aztecs great apprehension. They were a highly superstitious people who believed that coming events were foreshadowed by omens in nature. Aztec fortune-tellers were forever reading prophetic meaning into the smallest whisper of wind or rustle of leaves. No Aztec was more superstitious than the last emperor, Moctezuma II, better, though inaccurately, known today as Montezuma. Moctezuma worshipped the god Quetzalcoatl (pronounced "kwet-zal-ko-ah'-tull"), a kindly and pale-skinned deity who, Aztec priests said, had left Mexico ages earlier but had promised to return someday.

Legend had it that Quetzalcoatl would reappear out of the east, riding a magic raft of reeds across the ocean. He was expected to return in a "one-reed year," which occurred every fifty-two years in the Aztec calendar. Quetzalcoatl had failed to show up in the previous few one-reed years (1363, 1415, and 1467), but Moctezuma had high hopes for the next such year, 1519. The emperor even had a little speech ready to deliver if and when the god arrived:

> All this time I have been looking for you, waiting to see you appear from your secret place in the clouds and mist. The kings, my forbears, told me that you would return, [and] now it has come true.

As the one-reed year approached, Moctezuma saw what appeared to be a series of omens in the skies. Historian William Hickling Prescott, in his account of the invasion of the Aztec empire, reports what happened: "In [those] years, three comets were seen; and not long before the coming of the Spaniards a strange light broke forth in the east. It spread broadly at its base on the horizon, and rising in a pyramidal form tapered off as it approached the zenith. It resembled a vast sheet or flood of fire, emitting sparkles, or as an old writer expresses it, 'seemed thickly powdered with stars.'" This "strange light" was most likely the zodiacal light, a glow in the nighttime sky caused by sunlight reflecting off dust particles left behind by passing comets. So Moctezuma was seeing not only comets but also an eerie, comet-made glow in the heavens. These phenomena seem to have stirred dire imaginings in the emperor, for Prescott reports that Moctezuma thought he heard "low voices . . . in the air, and doleful wailings, as if to announce some strange, mysterious calamity! The Aztec monarch, terrified at the apparitions in the heavens, took counsel of . . . a great proficient in the subtle science of astrology. But the royal sage cast a deeper cloud over his spirit by reading in these prodigies the speedy downfall of the empire."

The final comet was the biggest and most portentous.

Moctezuma watched with growing wonder and apprehension as the comet grew brighter and brighter and then unexpectedly split in half. A great and unusual omen indeed —and an evil one, if the astrologer's forecast was accurate. Yet not all omens were menacing. Might not a comet's appearance just as easily signify a truly wonderful happening in the near future, such as the long-awaited return of Quetzalcoatl? The emperor clung to that hope as he watched the comet swing slowly across the sky. It moved steadily and deliberately, as if on some grave mission. Finally Moctezuma could bear to watch the comet no longer and retreated into the "Black Room" of his palace, where he was reputed to practice sorcery. The suspense was such a torture that his agitation turned into severe depression, and the comet preyed on Moctezuma's thoughts constantly.

Then along came Cortés. He had fair skin, appeared out of the east, and had crossed the ocean on a great raft, exactly as the legend had predicted; and by sheer luck his arrival coincided with the appearance of the comet. Moctezuma was overjoyed. The comet had been a good omen after all! Here was the noble fair deity, back at last! Under more ordinary, cometless circumstances, the Aztec ruler would probably have been on his guard, but his obsession with the celestial spectacle had blinded him to caution and to common sense. The comet had so mesmerized Moctezuma that he expected to see Quetzalcoatl, and Quetzalcoatl was what he saw. Moctezuma ordered no defenses raised against Cortés, for he could not believe his palefaced "divine" visitor might harbor any evil designs against him. Unfortunately for Moctezuma and his subjects, Cortés had anything but benevolent aims in mind, and the comet-addled Aztec leader was about to pay with his life for his blindness and naiveté.

Moctezuma went out to meet Cortés with all the groveling devotion of an altar boy meeting the pope. The emperor invited the Spaniards into the capital and gave them quarters directly across the plaza from his own palace. Cortés was impressed. He described the Aztec capital as "one of the most beautiful cities in the world" and reported that Aztec

civilization was "on almost a high a level as that in Spain
. . . well-organized and orderly. Considering that these people are barbarous, lacking knowledge of God and communi-
cation with other civilized nations, it is remarkable to see
all they have." Yet beauty and sophistication were not
enough to save the Aztec culture; it would have to be destroyed and its people subordinated to the Spanish empire if
Cortés were to succeed in his mission. Cortés decided the
easiest tactic was to grab the emperor and hold him hostage.
So trusting was Moctezuma, still bewitched by the comet,
that he never suspected Cortés's true nature and objective
until the Spaniards had taken him prisoner. Cortés demanded a huge ransom in gold and silver for Moctezuma.
The Aztecs complied, but even a fortune in precious metal
was too meager for the Spaniards' appetites. The conquista-
dors wanted still more. During a battle between the Spaniards and Aztec troops, a thrown rock hit Moctezuma on the
head and crushed in his skull. He died some hours later, a
victim of a comet and his own foolish faith in the occult. In
this case the old superstitions associated with comets came
true. The comet had presaged a war, the fall of an empire,
and the murder of a monarch.

COMETS, CHRIST, AND CAESAR

The ancient Romans lived in dread of comets. Like the Aztecs, the Romans were a superstitious people who saw in
every little event of nature a possible portent of things to
come. Therefore a comet's visit was a cause for alarm all
through the empire, and soothsayers did a land-office business as the fearful implored them to tell whether the comet
overhead would bring good luck or ill.

On several occasions comets played significant roles in
the lives of Rome's rulers. Julius Caesar had his career
bracketed by two bright comets, one of which appeared in
the year of his birth (100 B.C.), and the other, which passed

the earth just after his assassination (44 B.C.). That second comet played an important part in his deification, as the Roman biographer Suetonius tells us:

[Caesar] was fifty-five years old when he died, and his deification, formally decreed, was more than a mere official pronouncement . . . for on the first day of the games held by his successor Augustus in honor of [Caesar's] apotheosis, a comet appeared about an hour before sunset and shone for seven days in a row. The comet was held to be Caesar's soul rising to heaven; therefore [a star was] placed above the forehead of his divine image.

Thus a comet helped raise a dead monarch to immortality and link comets even more firmly in the public mind with the "changing of kingdoms." Another comet ushered Augustus out of existence in A.D. 14.

During the reign of Augustus, a memorable event took place in the skies. A bright object of some kind reportedly appeared in the east and moved along as if beckoning toward a spot in the oriental corner of the empire. This heavenly display is known today as the Star of Bethlehem and is an important part of the imagery of Christianity.

Historians still question exactly when the star appeared, but most estimates place it several years after the traditionally accepted date of Christ's birth. Astronomers debate what the star really was, and whether it even existed. If such a phenomenon did occur about that time, then there are several possible explanations:

A nova. *Nova* is the astronomical expression for an exploding star, which may suddenly increase hundreds of times in brightness before finally "burning out." (Science-fiction writer Arthur C. Clarke used this hypothesis as the basis for his famous short story "The Star," in which a spacefaring priest indicts God for wiping out a thriving alien civilization in a nova to announce the birth of Christ on earth.)

A conjunction. A conjunction is what happens when two or more planets come together at the same point in the skies. Jupiter and Saturn formed a conjunction in the constellation Pisces (the Fishes) about the time Christ is believed to have been born, and the two planets would have appeared to merge into a single, very bright, starlike source of light. British science writer David Hughes has argued persuasively for this explanation of the Star of Bethlehem, on the basis of its astrological significance. The zoroastrian priests of Babylon, from whence the Magi (the "kings" or "three wise men") came to worship the child, would most likely have read a powerful message into the conjunction of those two planets in Pisces. In ancient astrological thinking, Pisces was the ruling sign of Palestine, where Bethlehem was located, while Jupiter was said to stand for God's interests and activities, and Saturn for the doings of the Jews. A conjunction of Jupiter and Saturn in Pisces therefore signified that God had some extremely important business with the Jews in Palestine—in this case, the birth of the Messiah. (Elements of this astrological symbolism survive in the Christian churches today. Christ, for example, is often symbolized by a fish, the Piscean sign of his homeland.) If this is indeed the event that gave rise to the star of Bethlehem story, then it would make the most likely date of Christ's birth sometime in the early autumn of the year A.D. 7.

Yet the conjunction model of the Bethlehem event fails to account for another reported characteristic of the star: its motion. Matthew, the only Gospel author to mention the star in his writings, describes how it supposedly moved through the skies and led the Magi to Bethlehem: "And lo, the star, which they saw in the east, went before them, till it came and stood over where the young child was. When they saw the star, they rejoiced with exceeding great joy" (Matt. 2:9–10). In similar fashion, the sixth-century theologian Fulgentius wrote that the star "went always before the kings" and "was not fixed in the firmament," but rather was mobile, close to the ground. This is not the kind of motion one would expect to see from planets in the case of a conjunction. The two planets would come together, combine their lights into a single brilliant point source, and then part and go their separate ways.

Here we come to the third possible explanation:

A comet. This explanation would account for the brightness and apparent motion of the star. The problem here is that we have no reliable records of any comets at this particular moment in history. Yet the comet hypothesis remains a possibility, and indeed the Star of Bethlehem was portrayed as a comet by Florentine artists. Giotto's "Adoration of the Magi," painted about the year 1304, shows a clearly recognizable comet—probably modeled after Halley's Comet, which had appeared a few years earlier—passing over the manger. This appears to have been a truly spectacular appearance of the comet, for the Florentine historian Villani said it showed up in September of 1301, spewing "great trails of fumes," and was visible until early in 1302. Giotto's contemporaries also must have been deeply impressed by the sight of Halley's Comet, for many of them followed Giotto's example and depicted the Nativity star as a comet, though theirs were often stylized until they bore little resemblance to any comet ever seen.

Many leading philosophers of Giotto's time and earlier also interpreted the Star of Bethlehem to be a comet. The Genoese historian and theologian Jacobus de Veragine held that the star that led the Magi to the adoration at the manger, or the Epiphany (a word compounded from two Greek words, *epi* and *phanos,* which together mean "wonder in the skies"), was "a star newly created and made by God," a description that strongly indicates he was thinking of a comet. Learned men and women of that time commonly believed that God created new stars in the skies specially to commemorate major events on earth. The Byzantine scholar John of Damascus expounded this view when he wrote that comets "are not among the stars that were made in the beginning, but are formed . . . by divine command and again dissolved." Since this description fitted the apparition over Bethlehem, John figured it must have been a comet. John also pointed out how the motion of the star was most unstarlike:

> . . . sometimes its course was from east to west, and sometimes from north to south. [This is] quite out of harmony with the order and nature of the stars.

John's description suits a comet. His interpretation of the Nativity star also echoes that of the Christian theologian Origen, who in the third century A.D. speculated that "the star that was seen in the east we consider to have been a new star . . . such as comets, or those meteors that look like beams of wood, or beards, or wine jars." In Origen's time no distinction was made between comets and meteors, though we now know them to be two entirely different phenomena. Origen drew on the work of earlier Greek thinkers to justify his hypothesis, pointing out that if God would raise up new stars to announce a change of politics, such as the founding of a new regime or dynasty, then how much more justified God would be in creating a new star—a comet—to herald the birth of the Messiah! Origen admitted that no biblical prophecy referred specifically to the star over Bethlehem, but he thought he saw an oblique reference to that event in Moses' record of the words of Balaam, who said that "there shall arise a star out of Jacob and a man shall rise out of Israel" (Num. 24:17). Historian Roberta Olson, in an article for *Scientific American,* points out that the New English Bible translates the latter half of that passage as "a comet shall arise from Israel." Olson remarks that Origen "thereby contributed three important concepts to the legend of the star of Bethlehem. By repeating the Old Testament prophecy he implied a connection with the old tradition that the birth of prophets such as Abraham and Moses was heralded by a star, stressing the continuity between the two [Old and New] testaments. He clearly associated the star with a comet. And he stated unambiguously that comets could portend good."

There was also great evil associated with the apparition over Bethlehem. When King Herod saw the comet, he feared it might portend his downfall, and he asked an oracle to explain the meaning of the comet to him. The oracle confirmed his worst fears. The comet, she told Herod, announced the birth of a boy who would grow up to be a greater king than he. The king then resolved to eliminate his young rival, and ordered the Massacre of the Innocents. Herod's troops went out to slaughter all male infants born in Judea during the previous two years. To make sure the job was completed, Herod even decreed the execution of his own infant sons Aristobolus and Alexander, and his eldest son, Antipater. This bloodbath did not preserve King Herod's reign for long. Shortly after ordering this mass murder, he died, one chronicler reports, of a "loathsome disease." The object of Herod's wrath and fear, the infant Christ, survived. (Christ is only one of the prominent figures in religious history whose births were thought to have been announced by comets; the birth of the prophet Mohammed in A.D. 570 coincided with a comet's appearance as well.)

In A.D. 60 or 62 (the date is uncertain), a comet intruded on the business of one of Augustus' successors, Emperor Nero. Though the most famous tale of Nero's cruel and callous nature is probably a myth (rather than fiddling while Rome burned, the original manuscript of the story says he fretted), Nero nonetheless spilled a small ocean of blood during and after his rise to power and committed crimes that made his name a synonym for brutal abuse of authority. Far from trying to conceal his evil reputation, Nero seemed undisturbed when his foes accused him of murder and perversion, and he is said to have coined the saying, "Let them hate me, so long as they fear me!" When the comet appeared, however, Nero began to worry that fear might not restrain his enemies much longer, and they would have him assassinated. Maybe the comet was marking him for doom. What could he do to protect himself?

Nero's court astrologer Balbillus had an answer. If the comet portended the death of some prominent person, said Balbillus, that individual need not be the emperor; any outstanding personage would do. Therefore, in theory, Nero could simply decree the deaths of some notables, and the comet's work would be done for it, thus deflecting the peril away from Nero. That plan offered Nero both an illusion of safety and an excuse to get rid of his political opposition, so he took the astrologer's advice and ordered what one of Nero's biographers called a "wholesale massacre" of the nobility. Then he banished his victims' families from Rome for good measure. (Nero handed out death sentences for even the most trivial offenses; he condemned one nobleman to execution merely for wearing a sour expression in the emperor's presence.) But Nero's reprieve from doom was only temporary. Several years later, as another comet passed overhead, the tyrant was forced to flee his palace to avoid capture by his own mutinous troops, and finally committed suicide at a small villa outside Rome.

The emperor Vespasian, who reigned about a century after Nero, was a saner and more benevolent ruler, and he took a lighthearted view of the popular superstition about comets bringing death to heads of state. When he lay dying of a respiratory infection contracted on a march through the countryside in rainy weather, Vespasian joked about a comet that had recently appeared. "This hairy star is not coming for me," Vespasian said. "It must be meant for my foe, the King of the Parthians, because he has a full head of hair while I am bald." Vespasian's condition then took a sudden turn for the worse. His last recorded words were, "Dear me . . . I think I'm becoming a god." Thereupon he died, leaving one more anecdote in the lore of comets and death. (Increase Mather was fond of citing this anecdote in sermons as an instance of pagan blasphemy.)

Perhaps the most famous "victim" of a comet in European history was King Harold II of England. Halley's Comet made one of its periodic appearances in A.D. 1066, as Harold was fighting for his life and for his throne. England was about to be invaded by an army under the command of Duke William of Normandy, better known to succeeding ages as William the Conqueror. Looking up at the frightful apparition in the sky, King Harold must have wondered if he was doomed to defeat.

Indeed he was. A few weeks after the comet's passage, the Normans routed the defending English at the Battle of Hastings. Harold was killed when an arrow pierced his eye. William the Conqueror won the English throne, and the Norman Conquest became official. (Harold's defeat had more to do with faulty intelligence than with the comet. Norman troops wore their hair tonsured in the same fashion as monks, so that Harold's spies in Normandy mistook them for clergy and reported that William had recruited an army of priests to pray on his behalf! As a result Harold grossly underestimated the strength of William's forces and paid for that misjudgment with his life and his kingdom.) William's diminutive bride, Queen Matilda, had her husband's victory commemorated in the famous Bayeux Tapestry, a scroll of embroidered linen that tells in picture form—rather like a modern comic strip—the story of the Norman invasion and triumph. The comet appears in one panel as a highly stylized image, like a cross between a garden rake and a sunflower. Directly below, a group of men stand staring up at the comet. *ISTI MIRANT STELLA*, the Latin legend reads: "They are in awe of the star." So is King Harold II. The following panel of the tapestry shows him listening with a horrified expression as an aide rushes in to bring him news of the dreadful omen overhead.

Here are some more dire events that have been linked with comets:

COMETARY "OMENS" THROUGHOUT HISTORY

1949 B.C. Some historians claim that a comet's appearance coincided with the destruction of Sodom and Gomorrah.

1732 B.C. Arabian scholars recorded both a famine and the passage of a prominent comet.

1537 B.C. Chinese historians mentioned that a comet's appearance preceded a devastating flood.

1194 B.C. A fast-moving comet appeared in the Pleiades at about the same time as the fall of Troy.

431 B.C. The outbreak of the Peloponnesian War followed close upon the sighting of a comet.

184 B.C. Soothsayers warned the Carthaginian general Hannibal that a recently discovered comet meant he would die soon. This news depressed Hannibal so badly that he committed suicide.

63 B.C. King Mithridates of Pontus, in Asia Minor, received similarly bad news from his soothsayers and followed Hannibal's example.

A.D. 68 The historian Josephus records in his history of the Jewish Wars that the siege of Jerusalem was preceded by the appearance of "a comet shaped like a sword [that] was visible above the city for an entire year." Josephus rebuked his fellow Jews for ignoring such a plain warning from heaven. His account of the comet was used as evidence to support the view of Origen and other early Christian thinkers that a comet constituted the Star of Bethlehem. Some Jewish scholars interpreted Josephus' vision of the comet as a prophecy of the destruction of the Temple several years later. Josephus' account of the comet may also have given some ideas to Matthew, who wrote his Gospel shortly after the ruin of Jerusalem and scatters cometlike images here and there through his work.

A.D. 79 A comet was sighted in the skies over Italy. Several weeks later Mount Vesuvius erupted and buried the cities of Herculaneum and Pompeii under volcanic ash.

A.D. 312 Constantine saw "the sign of the cross" in the sky—most likely a comet—and was converted to Christianity.

A.D. 410 A comet was visible over Italy from April until early August. One week after the comet vanished, King Alaric of the Visigoths conquered Rome and put an end to the Roman Empire.

A.D. 451 A bright comet appeared days before the Goths halted the westward advance of Attila the Hun at the battle of Châlons-sur-Marne. Some 150,000 soldiers were reported killed. Attila died two years later, in the wake of another comet that was blamed for carrying off both him and his archenemy the Emperor Valentinian.

A.D. 531 Byzantine astronomers recorded the visit of a very bright comet, and soon afterward an earthquake struck Constantinople.

A.D. 684 A pestilence is said to have spread through the whole Eurasian land mass and killed millions of persons, shortly after a comet appeared early in the autumn.

A.D. 800 A comet coincided with the coronation of the Emperor Charlemagne.

A.D. 814 Another comet coincided with the death of Charlemagne, who on his deathbed thanked God for having sent the comet to warn him that his end was approaching.

A.D. 837 King Louis the Pious of France reportedly developed a religious mania at the sight of a comet that appeared in the Easter season. The king thought the comet was a warning from God, and Louis called in his court astrologer Eginard to see if the soothsayer could enlighten him. Eginard waffled at first. "Give me a little time," he asked the king, promising to deliver a full report on the comet's astrological significance the following morning. The king pressed him for an immediate answer and, when none was forthcoming, accused Eginard of trying to conceal sensitive information from him. Eginard tried to reassure the king that he had nothing to fear from the comet; did not the prophet Isaiah say, "Fear not signs in the heavens, like unto the heathen"? Eginard was no consolation to Louis, however, and the king went into a veritable orgy of prayers and devotions, ordering churches and shrines built to appease the imagined wrath of God.

A.D. 1000 Europeans grew more and more nervous as the end of the tenth century approached, for it was widely supposed that the world would end then, with Christ's return. Forewarned by a New Testament prophecy that "signs and wonders in the heavens" would immediately precede Christ's Second Coming, the public reacted with mass panic when a comet—described by one historian as resembling a "horrible serpent"—appeared in the "fateful year" 1000. The hysteria subsided the following year, only to flare up again when another comet showed up in 1002.

A.D. 1264 A comet blazed in the skies of the northern hemisphere for three months and was blamed for a plague in Italy and a war between the Russians and the Poles. Pope Urban IV died on the same evening that the comet faded from view.

A.D. 1303 Pope Boniface VIII interpreted a comet as a favorable omen for his continued reign. Several weeks later he was taken prisoner by King Philip of France and died of a stroke.

A.D. 1331 The birth of Tamerlane coincided with appearance of a comet over his birthplace in Turkestan. Tamerlane grew up to become one of the greatest warlords in history. His invasion of central Asia was accompanied by a comet in approximately the year 1382. Twenty years later another comet hung overhead as Tamerlane's troops took Constantinople. Tamerlane was leading an invasion of China when yet another comet allegedly whisked him away, in 1405.

A.D. 1347 A comet's appearance immediately preceded the outbreak of the Black Death, or bubonic plague, which killed a third of Europe's population.

A.D. 1456 An especially spectacular comet, described by one observer as having a "fan-shaped tail like that of a peacock [stretching across] half the sky," came along at a delicate juncture in European history. The Turkish army stood at the gates of Belgrade, and Pope Calixtus III was scared that a Mohammedan victory in eastern Europe might set off a domino effect that would put much of the rest of Europe under Mohammedan control. The last thing the Vatican needed was a monster comet hanging in the sky and demoralizing the faithful, so the pope allegedly ordered the ringing of church bells to drive away the comet, perhaps because he subscribed to the belief that one can scare off the devil by making a loud noise. One Vatican historian wrote:

A hairy and fiery star . . . made [an] appearance for several days, [and] mathematicians declared that there would follow . . . great calamity. Calixtus, to turn aside the wrath of God, ordered [prayers] that if evils were impending for humankind, [God] would turn them all upon the Turks, the enemies of Christendom. He also commanded . . . the bells to call the faithful at midday to help with their prayers those engaged in battle with the Turks.

On this occasion Calixtus is also said to have issued a bull against the comet, cursing it in the name of God and commanding it to go away. In theory the papal ban should have worked, because Scripture said that Christ had given Peter, founder of the Roman church, the power to bind and unleash phenomena in heaven; and it followed logically that the pope, as Peter's heir, had divine authority to send the comet packing. The pope's influence clearly did not extend into outer space, however, for the comet went on its slow and stately course, unperturbed by papal wrath.

A.D. 1572 The great comet of 1572 was interpreted as a sign of

God's anger over the slaughter of thirty thousand French Hugue-
nots in the St. Bartholemew Massacre on August 24 of that year.

1618. The outbreak of the Thirty Years' War followed closely on
the appearance of a reddish comet over Europe.

1755. A comet preceded the Great Lisbon Earthquake.

1835. Osecola, the leader of the Seminole tribe in Florida, saw an
appearance of Halley's Comet as an omen, and called on his people
to launch a war against white settlers. The Seminoles overwhelmed
the army garrison at Fort King and killed every last soldier. Osecola
personally scalped the fort's commander, General Wiley Thomp-
son. The Native Americans also attacked an army expedition in the
Florida swamps and killed all but a handful of the soldiers.

COMET SCARES

The "enlightenment" of the seventeenth and eighteenth
centuries dispelled many of the old superstitions about comets bringing calamities on the world—but not completely,
and not everywhere. Even in "civilized" Europe, the ancient bugaboos lingered in the public mind and could cause
mass hysteria at the mere mention of a comet's approach.

Consider the case of the great "noncomet" of 1857. This
international farce began when an anonymous German astrologer predicted that a comet would strike the earth on
June 13 of that year. Immediately, the comet became the
talk of Europe. The French astronomer Jacques Babinet
tried to reassure the public by stating that a collision between the earth and a comet would do no harm. He compared the impact to "a railway train coming into contact
with a fly."

The public, however, was not convinced. Fear of the comet
infected every level of French society, and the Paris corre-
spondent for the American journal *Harper's Weekly* wrote
that "women have miscarried; crops have been neglected;
wills have been made; comet-proof suits of clothing have
been invented; a cometary life insurance company (premi-
ums payable in advance) has been created . . . all because

an almanac-maker . . . thought proper to insert, under the week commencing June 13, 'About—this—time—expect—a—comet.' "

June 13 rolled around, and Europe braced itself for the expected shock. The *New York Daily Tribune* ran a brief editorial noting how modern humans, like their ancient ancestors, were making sacrifices to their gods in hopes of averting calamities from heaven:

> This is the day set down for the appearance of a tremendous comet, and the utter annihilation of the world thereby. In Europe, and even in this country, not a little real apprehension has been manifested, chiefly, of course, among the ignorant and superstitious. We are assured that a great many servant girls, and others of the illiterate, have been in a state of painful alarm, and have to an unparallelled degree sought the consolation of their spiritual advisers; some of them even going so far as to make a gift to the Church of the little hoard that frugality had amassed by hard service. After this substantial evidence of faith, they will doubtless feel their disappointment keenly; indeed, one could almost wish—if it were not for personal consequences—that they might be justified in their credulity.

June 14 dawned, no comet arrived, and the world was still intact. An astrologer's "prophecy" had frightened millions, in the total absence of danger.

Such comet scares occurred frequently in the 1800's. The return of Biela's Comet in the autumn of 1832 led to a prediction that the comet would collide with the earth and cause tremendous destruction. But in fact there was no risk of a collision. The comet did intersect the earth's orbit, but at a point some 60 million miles from the earth's position on that day. Forty years later the British journal *Nature* reported that in another comet scare "many weak people have been alarmed, and many still weaker people made positively ill, by an announcement which has appeared in

almost all the newspapers, to the effect that Prof. Planta-
mour, of Geneva, has discovered a comet of immense size,
which is to 'collide,' as our American friends would say,
with our planet on the 12th of August next." In a pattern
that would become familiar over the next few decades, the
scientific journals pointed out how the comet—a mere
"wind-bag," *Nature* called it—posed virtually no threat to
the earth because of its extremely low density; the chance of
the earth colliding with a solid part of a comet is roughly
equivalent to a person bumping into one particular bit of
pollen during a spring stroll in Central Park. The scientists
might as well have saved their time and effort, however, for
reasoned explanations of the cometary phenomenon did
nothing to diminish public fear. Ordinary citizens attributed all manner of awful powers to the comet, including
Jove's capability of throwing lightning bolts. During one
nineteenth-century comet scare in Paris, families sat neck-
deep in the waters of the Seine in an effort to escape electro-
cution.

Panics like these inspired nineteenth-century French
science-fiction writer Jules Verne to poke gentle fun at the
public obsession with comets. In his 1877 novel *Hector
Servadac (Off on a Comet),* a comet collides with the earth
and shatters the planet to bits. A group of survivors, drifting through space on a small chunk of the earth, at first
think their fragment is merely an island. Their attitude
changes after they encounter the uprooted Rock of Gibraltar floating through the void with two stiff-upper-lip Brit-
ons aboard, playing chess and pretending that nothing has
happened. Verne continued to spoof comet scares in a sequel
entitled *To the Sun!* Here Verne describes how an imaginary comet grabs up a chunk of North Africa and takes it
and its inhabitants—a typical assortment of Verne characters including a square-jawed hero, a comic noncom, and a
crusty old professor—on a tour of the solar system, then deposits them back on earth so gently that no one believes
they have been gone at all, much less journeyed deep into
space. In fact, no comet could behave as those in Verne's

novels do; but Verne never let the laws of physics stand in the way of a good story.

Good humor was conspicuously lacking from *Ragnarok,* a bizarre book on comets published in 1882 by the Populist Irish-American politician and best-selling author Ignatius Donnelly. A gray-haired man with a walrus mustache and a slightly wild look in his eyes, Donnelly was the Populist party's candidate for vice-president of the United States in 1900, running on a reformist ticket that newspapers dubbed the "Platform of Lunacy."

Donnelly had several notably odd convictions. He believed that the lost land of Atlantis had really existed and that the plays of Shakespeare contained a code that proved Francis Bacon had written them. He also believed that a giant comet had passed close to the earth ages earlier and had caused all manner of natural calamities. Intense heat from the comet, Donnelly said, set off huge fires that raged across the face of the world. He added that the comet had dumped vast amounts of dust upon the earth. According to Donnelly, this comet-caused dust storm triggered earthquakes, leveled mountains, and initiated an ice age. Donnelly used his imaginary comet to account for some of the miracles described in the Old Testament. He proposed that "even that marvelous event, so much mocked by modern thought, the standing-still of the sun at the command of Joshua, may be, after all, a reminiscence of the catastrophe." He cited Native American legends according to which the sun stood still, and Donnelly encouraged the reader's imagination to run wild when he wrote: "Who shall say what circumstances accompanied an event great enough to crack the globe itself into immense fissures?"

Donnelly's "evidence" included the biblical story of the trials of Job, which Donnelly thought to have been based on memories of the comet's devastation. Donnelly thought he had more concrete evidence in the deposits of till—thick layers of mixed clay and gravel—found in his home state of Minnesota. Orthodox geologists said the unconsolidated sediment was dropped there by glaciers. Donnelly claimed that till was part of the dust showered on the earth by his hypothetical comet. Donnelly's "scientific" evidence was rubbish, but he knew how to tell a good story, and *Ragnarok* sold well. Readers thrilled to his descriptions of the "glaring and burning monster" in the sky baking the world with its "unearthly heat" and shaking the globe with "thunders beyond all thunders."

Donnelly was a politician as well as a storyteller, and could not resist inserting a Populist sermon at the end of his comet yarn. He urged the wealthy reader to "readjust the values of labor [so] that plenty and happiness, light and hope, may dwell in every heart. . . . from such a world God will fend off the comets with his great right arm, and the angels will exult over it in heaven."

Scientists did not exult over Donnelly's book. They considered his fictional comet a piece of quackery and sensationalism. The scientific community reacted in much the same fashion when, in 1893, public fear of—and fascination with—comets inspired a French science writer named Camille Flammarion to pen a best-selling disaster story called *La Fin du Monde (The End of the World)*, telling of the effects of a fictional collision between the earth and a comet fifty times its size. Published in 1894, the story was an immediate sensation and later was filmed by the French movie pioneer Abel Gance, who conjured up frightening images of the people of Paris going mad under the influence of psychoactive gases in the comet's tail. Flammarion described his make-believe comet in lurid prose:

[The] Comet dominated the world—a scarlet ball with jets of yellow and green fire . . . the flaming rays would descend upon the earth. . . .

The dryness of the air became unbearable. Heat like that of a great burning oven came from overhead. A horrid stink of burning sulfur—doubtless due to electrified ozone—poisoned the air. . . .

Everyone then saw their time had come. . . . "We are on fire!" they cried.

In the novel, poison gases from the comet pervade the atmosphere and make breathing impossible. The vast majority of the earth's inhabitants perish. Only a small fraction of the population survives, by descending into hermetically sealed underground vaults. The last person to go underground is a young woman from California whose nerves have been toughened by earthquakes there. She pauses for a last look around and sees a giant "white-hot meteorite rushing southward at the speed of lightning." The multitudes left behind on the surface are overcome by heat, noise, earthquakes, and toxic gases. "Laid low like dumb animals, they met their doom," wrote Flammarion. "The end of the world had come."

Flammarion was probably writing in jest, for he was a widely published science writer and must have been aware that the toxic substances in a comet's tail are much too rarefied to do anyone harm, even if the earth passed directly through the tail. He must also have been aware that no comet has anything like fifty times the mass or volume of the earth; the true proportions are closer to the other way around. Yet his story scared the wits out of many readers and was still a vivid memory when Halley's Comet came around for its next appearance in 1910. This time Flammarion's vision of an asphyxiated world seemed more terrifying than ever, for the earth was expected to pass directly through the comet's tail. Here is one contemporary account of the perceived threat from the comet's chemistry:

Our atmosphere contains a certain amount of hydrogen, a marvelously light gas . . . characterized by an extreme inflammability. . . . If the gas shall ever be touched off by this [comet], our planet will be ignited. The whole atmosphere will become a seething ocean of flame, in which forests and cities will burn like straw, in which oceans will boil away. . . .

Fears of a comet igniting the atmosphere were of course groundless. There is virtually no free hydrogen in our air, because the slightest spark makes it combine with free oxygen to form water (H_2O). All the same, public worry increased to the point where *Scientific American,* in a 1909 article about the comet, addressed the issue of comet-earth encounters:

> On May 18th next the earth will be plunged into the tail of Halley's Comet, and the head of that body will be but 15,000,000 miles away. It is but natural that a thinking man should ask: Is there a possibility that the earth may encounter a comet and thus come to a frightful end?

The magazine reassured its readers that the world had nothing to fear from Halley's Comet on this visit, and specifically denounced Flammarion's wild vision of global asphyxia:

> It is true that a comet's tail is composed of poisonous and asphyxiating . . . vapors . . . but it is also true that the actual amount of toxic vapor is so small that when the earth is brushed by the tail of Halley's Comet, the composition of the atmosphere will not be so affected that a chemist could detect it. Flammarion has shown us terrified humanity gasping for breath in its death struggle with carbon monoxide gas, killed off with merciful swiftness by cyanogen, and dancing joyously to an anaesthetic death, produced by the conversion of the atmosphere into nitrous oxide or dentist's "laughing gas." No one of any common sense should be alarmed by these nightmares. . . .

Despite such comforting words, "cometophobia" spread like wildfire through the populace and was fanned by soothsayers who wanted a bigger piece of the action. Historian Edwin Emerson, in his lively 1910 pamphlet *Comet Lore,* quotes the French psychic and astrologer Mme. de Thebes as issuing a gloomy forecast for the year 1910: "The earth is under a terrific strain from comets and planetary revolutions. Human destiny is red. That means blood. Political events are black. Terrible changes are imminent."

Mme. de Thebes had especially dark tidings for the United States. "The strain of the stars," she warned, "will be most severely felt in America. The people of America will have to pay dearly for all their riches and sudden prosperity. With the coming of another comet disaster will descend upon America."

Like most psychics then and now, Mme. de Thebes was hard pressed to come up with specific details of the woes about to be visited on America. "I dare not say all that is revealed to me," she said gloomily. "It would be too terrible." She did predict a "financial crash" followed by numerous suicides, and in fact the famous stock market crash did send a few ruined ex-millionaires hurtling out of windows on Wall Street; but that calamity took place in 1929, almost two decades after Halley's Comet inspired the French seeress's forecast, and so the accuracy of Mme. de Thebes's predictions is suspect at best.

One American, at least, was not worried about Mme. de Thebes and her warnings of doom. Author Mark Twain viewed the comet's return with happy anticipation. He explained his reasons thus, in 1909: "I came in with Halley's Comet in 1835. It is coming again next year, and I expect to go out with it. It will be the greatest disappointment of my life if I don't go out with Halley's Comet. The Almighty has said, no doubt: 'Now here are these two unaccountable freaks; they came in together, they must go out together.' Oh, I am looking forward to that." His wish was granted. He died in 1910.

Shortly after the comet was sighted that year, General William Booth, founder of the Salvation Army, confessed his apprehensions in a speech delivered in London. "We are this year approaching the end of all things," said Booth, and warned that the comet's visit would be accompanied by events "far surpassing in horror any disaster that has gone before. All things will be wound up. Besides a deluge of water sweeping parts of the world and its inhabitants, there

will be fierce destruction by fire." Booth's remarks may
have been inspired by a prominent French astronomer who
was quoted as saying that the comet's tail might "upset the
atmosphere of the earth, causing rains of long duration, and
consequent floods and overflows of rivers."

United States newspapers covered the comet's approach
in detail. Mary Proctor in the *New York Times* wrote a set of
serious articles about it and mentioned in one of them the
"peculiar red glare" of the comet when seen low in the sky.
The comet's "ruddy hue" was caused, she wrote, by "the
light of . . . the comet gleaming through the mist low down
on the horizon." She mentioned that the comet's red color
had caused a scare on Bermuda, and she admitted the fear
was understandable: "To the imaginative spectator it would
have suggested an ill-omen in the sky, for it was an awe-
inspiring spectacle. . . ."

Imaginative spectators there were aplenty, and many of
them saw the comet as an "ill omen" indeed. Here are some
examples from news stories printed in May 1910:

One Blanche Covington of Chicago decided to kill herself before the
comet had a chance to asphyxiate her. So she locked herself in her
room and turned on the gas. Fortunately for her, a neighbor called
the police, and Covington was rescued. The *San Francisco Bulletin*
reported another suicide attempt:

> Denver, Colo., May 19—Driven to despair by brooding over the
> possibility of the annihilation of the world by Halley's Comet, Mrs.
> Jeannette Neibert, 31 years old, yesterday attempted to kill her-
> self by swallowing morphine. Little hope is entertained for her re-
> covery.
> According to the woman's husband, Mrs. Neibert . . . thought
> of little except the comet for the last few days, and declared that
> she was crushed by a premonition of disaster. Her last words be-
> fore sinking into unconsciousness after taking the poison were,
> "I think the comet—".

San Franciscans seemed to take the comet's appearance more
calmly. In another story on the same page, the *Bulletin* said that

"San Francisco is doing very well, thank you," and added: "Beyond a little nervousness on the part of the gentler sex, fear of the comet has not manifested itself in any way." The newspaper admitted that the comet might be dropping "foolish powders" from its tail onto other areas, however, because of some precautions being taken against the expected threat from the comet's tail. "A man at Pasadena . . . dug a cave in a hillside, equipped it with a . . . gas-proof door, went inside, and pulled the hole in after him," the *Bulletin* reported.

The fall of "foolish powders" was not confined to the West. Children at Public School 110 in New York panicked when a flash picture was taken of them. They thought the burst of light came from the comet and marked the end of the world.

The fear of death fell over the Deep South as well. One newspaper reported from Lexington, Kentucky: "Certain that the approach of Halley's Comet spells their doom, many farmers in Kentucky and scores of [the] ignorant . . . are preparing for the end of the world. . . . For the last week they have been holding 'union services,' which have been crowded."

The Korean government feared the comet signified that their country would soon fall to the invading Japanese. Vice-Minister Yi Feng Lai fled Seoul in hopes of escaping Korea's imminent collapse. The Japanese were equally worried about the comet and reinforced their garrisons in Korea to repel attacks from Korean fanatics who were using the "omen" overhead to rally the people in an uprising against the invaders.

A twenty-pound blade came loose from a cooling fan at a motor car company on New York's Eighth Avenue, went flying through the air, and landed on a passing trolley car, shattering all the windows. "The comet!" screamed the passengers as they ran from the car.

Actress Olga Nethersole was worried not about her own safety, but rather about her pet Mexican hairless dog. "If the comet touched her, my poor pet has no hair to protect herself," Nethersole explained.

A voodoo doctor in Haiti was reported selling "comet pills" at one dollar a box, for protection against the nonexistent threat from the comet's "tail fumes."

The comet terrified the Mexican peasants along the Rio Grande, and the *San Francisco Call* printed a news report from El Paso: "Hundreds of Mexicans from the villages along the Mexican border gathered around crosses erected on the hills . . . awaiting the appearance of the fiery comet they believe is hurrying to destroy the world." The story added that "hundreds [more] have sought refuge in caves and canyons in the mountains."

The Plaza Hotel in New York introduced a "comet cocktail." The ingredients were a secret but were said to induce a "feeling of mental buoyancy."

The following classified ad appeared in a newspaper in Johannesburg, South Africa: "Gentleman having secured several cylinders of oxygen and having bricked up a capacious room wishes to meet others who would share the expense for Wednesday night [when the earth was expected to pass through the comet's tail]. Numbers strictly limited."

One guest at the Waldorf-Astoria left for the Adirondacks as the comet approached. She explained that she would rather face the destruction of the world at her mountain lodge than in New York.

Such reactions to the comet amused many New Yorkers, especially the members of the Thirteen Club, an antisuperstition society that held a merry dinner on Friday, May 13. One hundred and sixty-nine members of the club were present, seated in groups of thirteen at thirteen tables, on which everyone was careful to spill salt. Some in the audience sat under open umbrellas while listening to the guest speaker, Columbia University astronomer B. A. Mitchell, warn in mock-solemn tones of the peril from the comet. Mark Twain and Britain's King Edward had just passed away, and Mitchell had advice for those who thought they might be next. "There is no doubt in the world," said Mitchell, "that the comet's tail contains cyanogen gas, found in prussic acid, a poison so powerful that a drop on the tongue would cause instant death. . . . Very probably some of you would like to take to a dugout or subcellar that day. If you are really so afraid that you are all going to die on that day, I hope you will make your bank accounts over to me, as I am willing to take my chances."

As the world watched, the comet put on a splendid show, conducting itself in a "most whimsical fashion," as one scientific journal put it. At one point it shed its tail entirely and left its detached train of gas and dust hanging in outer space, suddenly headless, while its head continued on its way sunward. The tail then grew back in a three-pronged configuration, even longer and more handsome than before. The tail made a glorious target for photographers, and a quest for pictures of Halley's Comet very nearly cost two balloonists their lives. They set off by air from Quincy, Illinois, and took photographs of the comet from an altitude of eighteen thousand feet. Then they began their descent toward a landing in Kentucky. A malfunction caused their balloon to collapse just prior to touchdown, however, and the aeronauts fell the last three hundred feet to the ground. They were saved by an air mattress in the bottom of the balloon's gondola.

At big-city hotels, guests started organizing "comet parties." Astronomer Ralph Grace hosted one such gathering on the roof of New York's Hotel Gotham, complete with hanging Japanese lanterns carefully arranged so that their glow did not interfere with comet viewing. A Hungarian orchestra played popular songs, including one called "A Trip to Mars." Grace handed out small souvenir telescopes to his guests. While New Yorkers panicked or partied, Berliners were enjoying "comet picnics" under the stars, and in rural North Carolina much of the work force vanished into the churches for days on end, to pray for deliverance from the horror overhead. In some towns it was impossible to hire laborers. "No more paydays," the workers said gloomily. Physicists were taking a less morbid interest in the comet's encounter with the earth; one Norwegian scientist trekked all the way to the northernmost tip of Scandinavia to wait and see how it might affect the northern lights. Meanwhile the U.S. Hydrographic Office in Washington, D.C., instructed wireless operators to keep a close eye on their equipment during the comet's passage and see if anything strange happened to cable communications.

Finally came the fateful night when the comet's tail was supposed to brush the earth. The world's population held its breath, in some cases literally, while the planet passed through the comet's tenuous "exhaust."

Nothing happened. The air remained nontoxic. Wireless signals zipped through it unhindered, and the Norwegian professor came home with no results to show for his travels. Humankind heaved a collective sigh of relief, and life returned to normal. The astronomers tallied up their observations of the comet and developed their photographs. The soothsayers counted their take and began looking for new ways to fleece a gullible public (*"Mundus vult decepi,"* remarked Edwin Emerson—"The world wants to be deceived"), while the earth continued on its way through space as if nothing unusual had taken place at all.

In fact, nothing had. Halley's Comet of 1910 was merely one of millions of comets that have passed by the earth in ages past and will continue to do so for ages to come, whatever superstitious humans may think and say and do.

The Comet
Hall of Fame

A NOTE ON NOMENCLATURE

Halley's Comet may be the most famous comet of all, but there are many other well-known comets, and they may go by a variety of names. Officially a comet is designated by the year of its discovery, followed by a letter or Roman numeral signifying the order in which the comet was discovered that year. For example, Halley's Comet, on its 1910 appearance, was catalogued as comet 1910 II or 1910b, because it was the second comet discovered in 1910. Comets are also commonly named after their discoverers, such as Comet Smith and Comet Jones. It is not uncommon for two or more astronomers to discover a comet simultaneously, and when this happens the comet may take on a double or even triple name like Comet Smith-Jones-Brown. Older and more familiar comets are often labeled with their discoverer's name first: e.g., Smith's Comet. The following list reflects common usage.

Comets are not always named for the person who discovers them. Encke's Comet, for example, was discovered by another astronomer but was named after Johann Encke because he was the first to recognize the periodic character of its orbit.

To avoid confusion, astronomers designate some comets

with a *P* prefix if they have brief periods and return every few years. Instead of writing down Jones's Comet as 1910 I, 1915 II, 1920 IV, and so on, it is easier to call it "P/Jones."

The references here to "degrees of arc" may require some explanation. Astronomers divide the circumference of the heavenly sphere into 360 degrees. A comet, measured from head to tail, appears to span an arc across the sky, and so its length is gauged in degrees of arc. A typical comet might span 20 or 30 degrees of arc, whereas a giant comet might stretch across 60 degrees (one-sixth of the way around the heavenly sphere) or even more.

SOME FAMOUS COMETS

Comet Arend-Roland (1957 III). This comet showed up on an observatory photograph and is remembered for its prominent antitail, a jet of material expelled toward the sun, which extended out some 15 degrees of arc from the comet's head.

Bennett's Comet (1970 II). A South African astronomer named John Bennett discovered this comet (Fig. 2-1) in the vicinity of the Magellanic Clouds, two "satellite galaxies" that orbit around our own galaxy and are visible only from the southern hemisphere. Bennett's Comet displayed gas tails and dust tails and glowed a pale yellow. The nucleus was unusually active and appeared to be spinning rapidly, for it was surrounded by a "pinwheel" structure like that of a whirling firework.

Biela's Comet (P/Biela). In 1772 two French astronomers noticed a very faint comet and recorded its appearance. They probably had no idea that this comet would become one of the most famous in history. In 1801, Friedrich Wilhelm Bessel, the director of the Albertus University observatory in Konigsberg, East Prussia, rediscovered it and, noting its similarity to the comet of 1772, concluded

Figure 2-1. Bennett's Comet. A yellowish comet with a highly active nucleus, Bennett's Comet was discovered in 1970. (D. Wereb)

that the two were actually a single periodic comet. Later Bessel decided that his observational data had been inaccurate, and he published a retraction. The comet he saw in 1801, he announced, was not the same as the comet seen twenty-nine years earlier. Bessel's retraction was in error, however, for an amateur Austrian astronomer named Joseph Morstadt reviewed Bessel's work, checked his calculations carefully, and decided that Bessel had been right the first time: the two comets were one and the same. Morstadt figured that the comet would return again early in 1826, and he was on the watch for it when that year rolled around.

To make sure the comet did not slip by unnoticed, as it appeared to have done on its visits in 1813 and 1819, Morstadt called on his friend Captain Wilhelm von Biela for help. Biela commanded an army garrison and ordered all the soldiers on watch to keep their eyes peeled for comets. The soldiers' vigilance paid off, for on February 27, 1826, one of them spotted the comet. Biela watched it for about three months, used his observations to calculate its orbit, and figured that it circled the sun once every six years and nine months. Morstadt and Biela were jubilant, and Bessel's initial hunch was vindicated.

The comet was due again in 1832. Italian astronomer Giovanni Santini predicted that the comet would be back late in autumn of that year and would reach its nearest approach to the sun on November 27. Santini was off by only twelve hours. Santini predicted the next appearances of the comet (now known as Biela's Comet) for July of 1839 and February of 1846. Biela's Comet was not visible in 1839 because it did not appear in the nighttime sky and was too dim to be seen in the daytime; but it more than made up for that unspectacular showing in 1846, when it astounded the scientific community by splitting in half before the very eyes of astronomers.

The breakup of Biela's Comet provided American astronomers with a chance to upstage one of their British colleagues. The bumbling British astronomer James Challis observed the fission of Biela's Comet but simply could not believe what he was seeing, so he hesitated to report his observation. Astronomers at Yale University and the U.S. Naval Observatory in Washington, D.C., also observed the comet's disintegration and hurried to report their finding, thus beating Challis to the credit for it. Challis tried to explain his blunder by saying he had actually been searching for the planet Neptune, which had been discovered earlier that year. When he missed finding Neptune, even though he had been told precisely where to look for the planet, he tried to excuse himself by claiming he had been looking for the comet instead.

The two "Bielid" comets came around again in 1852 and were observed by Father Angelo Secchi of the Vatican Observatory. On this approach the comets were about one and a half million miles apart and were indistinguishable from each other, so it was impossible to say which, if either, was the parent fragment. This was the last time the twin comets were ever seen. Soon afterward the comets broke up and dissipated.

Donati's Comet (1858 VI). One of the most spectacular of the nineteenth century, Donati's Comet displayed a broad dust tail along with two narrow pencil-like gas tails and was among the first comets to be studied by spectroscope (see Chapter 5). The dust tail measured an astonishing 40 degrees—one-ninth of the circumference of the heavens—and one of the gas tails was even longer. Fortunately for European and American astronomers, Donati's Comet was clearly visible from the northern hemisphere.

Comet Humason. This inconspicuous comet (1962 VIII) is interesting for several reasons. For one thing, it is the only comet named for a mule skinner. Milton Humason, its discoverer, led pack mules up and down Mount Wilson in California to carry materials for the construction of the observatory there. When he fell in love with a young woman whose father frowned on his humble profession, Humason determined to improve himself, and he took a variety of jobs at the observatory. According to one story, an assistant astronomer fell sick one evening and Humason was named to take his post at the telescope. Humason did the job so expertly that the observatory staff enlisted his help in further observations, and he became a friend and colleague of the distinguished astronomer Edwin Hubble. Comet Humason is very faint as comets go, because its perihelion, the closest point in its orbit to the sun, is about four astronomical units, or roughly 375 million miles. Consequently, it never comes near enough to the sun to develop a splendid tail like that of

Halley's Comet. That of Comet Humason looks short and ragged, contains virtually no dust, and has a blue tinge created by emissions from ionized hydroxl, cyanogen, and other gases.

Comet Ikeya (1963 I). Discovered by the young Japanese amateur astronomer Kaoru Ikeya, Comet Ikeya has an inspiring story associated with it. Ikeya was the son of an unsuccessful businessman who had sunk into alcoholism and brought shame to his family—an awful curse in status-conscious Japan. The unskilled boy wanted to do something that would bring honor and glory to his family and restore its tarnished name. He decided to hunt for comets. Discovering a new comet, he thought, would make up for his clan's disgrace. He was right: With his homemade telescope, he soon spotted a fuzzy patch of light that turned out to be Comet 1963 I, or Comet Ikeya, and the Ikeya family was suddenly world-famous.

Comet Ikeya-Seki (1965 VIII). Ikeya shared credit for the discovery of this bright comet with a Japanese guitar teacher named Tsutomu Seki. They found it simultaneously on the morning of September 18. A storm had swept away air pollution and clouds the previous evening, so that conditions were perfect for viewing. Seki said later that the comet glowed "like a street light on a foggy night." It soon grew much brighter than anyone had anticipated and became one of the most closely studied comets of our century. Ikeya-Seki was a sun-grazing comet and gave astronomers a valuable opportunity to observe such a comet as it passed perihelion. It came so close to the sun that at one point it looked as if the two would collide; but Ikeya-Seki cleared the sun by about a third of a million miles, or slightly more than the distance between the earth and the moon. This close approach inspired some preposterous tales in the mass media. One ill-informed commentator claimed the comet would pivot across the whole sky like the beam of a giant searchlight, when in fact the comet's tail was much less extensive. At this time Ikeya-Seki could be seen plainly in the moments just before sunrise, when it had cleared the horizon but the sun had not. Unfortunately for amateur comet watchers, it happened along at a time when much of the world was undergoing a spell of stormy weather that obscured the skies and precluded viewing for much of the comet's visit. Just before perihelion it glowed as brightly as the full moon. The fierce heat and particle output of the sun at that close range blasted away most of the comet's tail, but Ikeya-Seki quickly grew a new tail much longer and more beautiful than the one it had before rounding the sun. Before long the comet could be seen in daylight with the unaided eye. Big and beautiful though it was, the media virtually ignored it after perihelion because Ikeya-Seki had been something of a letdown before it grazed the sun. That error cost the newspapers and TV some glorious pictures, for Ikeya-Seki after perihelion really did have the "searchlight" appearance that had been expected earlier. The comet's tail grew shorter and less brilliant as the days passed, and by mid-January Ikeya-Seki had vanished into the darkness of outer space.

King David's Comet. Sometimes identified as a return of Halley's Comet, "King David's Comet" is mentioned in the first book of Chronicles in the Old Testament. Though the Bible never uses the word *comet* to describe the phenomenon, it could hardly be anything else:

And David lifted up his eyes and saw the angel of the Lord standing between the earth and heaven, having a drawn sword in his hand stretched out over Jerusalem. (I Chron. 21:16)

The Bible says God sent this apparition to chastise Israel in general and King David in particular, for the king had been abusing his authority of late and needed to be shocked into repentance. Evidently the warning worked, for David saw the error of his ways and turned back to righteousness.

Note the similarity between this passage and the description of the comet seen over Jerusalem by Josephus in A.D. 66. Apparently King David's Comet was extremely bright, for the Bible report indicates that the "angel" could be seen in daylight. Indeed, the heavenly "visitation" appears to have been so bright that some comet experts doubt it was a comet at all; but the sword metaphor indicates that something with a scimitar shape hung in the sky over Israel about then, and hardly anything but a comet fits that description. Whether or not Halley's Comet was responsible for this incident from Scripture, no one knows for sure; it was due for a return about the time of King David's repentance (somewhere around 1000 B.C.), but some other bright comet is also a possible explanation.

Comet Kohoutek (1973 XII). Very few celestial events of recent years have generated so much publicity—and ended in such widespread disappointment—as the appearance of Comet Kohoutek in late 1973. Dr. Lubos Kohoutek of the Hamburg Observatory discovered the comet's image on a photographic plate taken in March. At that time it was merely a minuscule blob of light with no remarkable characteristics. The comet's orbit was calculated soon afterward, and it was determined that it would be a sun-grazer, passing within ten million miles of the sun at perihelion. Since many sun-grazers turn out to be huge and brilliant comets, the mass media began to advertise Kohoutek as the most spectacular of the century and to predict that its tail would reach all the way across the sky during the comet's closest approach, around Christmas. "Comet fever" infected much of the world as Kohoutek drew near. At least one religious organization issued a pamphlet denouncing the comet as the "Christmas monster." The American journal *Christianity Today* pointed to the comet as a sign of the expected return of Christ, made reference to the "signs and wonders" that the New Testament prophesied as heralds of Christ's second advent, and crowed that "the King is coming!"

Observers who expected a truly colossal comet were let down, for Kohoutek, while an attractive sight and a highly revealing study for astronomers, attained nothing like the size and brightness widely predicted for it. About the time of its closest approach to the earth, one well-known American astronomer hosted a party at which he served a bland alcohol-free punch: "A fake punch," he told his guests, "for a fake comet." Yet Kohoutek was by no means inconspicuous. Its tail reached a length of about 15 degrees. Kohoutek's passage coincided with a United States Skylab space mission and astronauts had an opportunity to watch the comet closely. Among the features they observed was an antitail that appeared to be made up of particles about a millimeter in diameter, or approximately the size of grains of sand. Some observers also thought they detected evidence of an icy "halo" surrounding the nucleus of the comet.

Spectroscopic observations of Kohoutek yielded some interesting discoveries. For one thing, gas molecules in the comet's tail appeared to be "soaking up" sunlight on the comet's approach toward the sun, and then letting that solar energy "leak out" again as light emissions after passing perihelion. Also, Kohoutek appeared to be radiating energy in radio frequencies—a curious event that was observed one day only. Exactly what caused the radio emissions is a mystery. Perhaps the most impressive discovery concerning Comet Kohoutek was the size of its surrounding hydrogen cloud. For a while that cloud made the comet the largest single object in our solar system in terms of volume. Comet Kohoutek was a long-period comet with an orbit thousands of years in duration. This was most likely its first visit to the inner solar system.

Comet Mitchell-Jones-Gerber (1967 VII). One of the astronomical highlights of 1967, this comet had an impressive tail and was a beautiful sight through binoculars.

Comet Mrkos (1957 V). This comet "sneaked up" on the earth and was not detected until it was very close to perihelion. Its tail reached a maximum length of about 15 degrees. For a while the comet displayed a prominent antitail.

Comet 1910 I. This big, bright comet goes by several names, including one that it acquired by accident. It was spotted by miners in the Transvaal of South Africa in January. They reported their discovery of a "great comet" by telegram, but the message was garbled in transmission, and the comet was mistakenly dubbed Drake's Comet. It is also known as the Miners' Comet and is said to have caused widespread panic in the Iberian peninsula. The comet had a long, narrow tail of gas and a very luminous dust tail. According to one report, the tails spanned more than 40 degrees of arc. An eyewitness account conveys something of the comet's appearance: "The comet . . . was a gorgeous object. The picture presented in the western sky will never be forgotten. . . . High up in the southwest shone . . . the great comet itself, shining with a fiery golden light, its great tail stretching some seven or eight degrees above it. The tail was beautifully curved like a scimitar, and dwindled away into tenuity so that one could not see exactly where it ended. The nucleus was very bright, and seemed to vary. . . . The tail, too, seemed to pulsate rapidly from the finest veil possible to a sheaf of fiery mist." The Miners' Comet might have gone down in history as the most spectacular comet of that year had it not been upstaged by Comet 1910 II, better known as Halley's.

Comet Seki-Lines (1962 III). A very brilliant comet, Seki-Lines had a prominent "nuclear shadow" that ran down the middle of its tail and appeared to divide the tail in two.

Comet Tago-Sato-Kosaka (1969 IX). Another Japanese discovery, Comet Tago-Sato-Kosaka looked much like the 1957 appearance of Comet Mrkos but was remarkable for a prominent "flare" or outburst of material from the nucleus.

Tebbutt's Comet (1861 II). Discovered in the spring of 1861 by John Tebbutt of New South Wales, Australia, Tebbutt's Comet was first spotted through a sextant and reportedly attained a length of 120 degrees, though that figure may be an exaggeration.

Figure 2-2. Comet West. One of the most spectacular comets ever seen, Comet West was virtually unnoticed by the public on its 1976 visit, because Comet Kahoutek in 1973 had turned out to be much less glorious than expected, and so the mass media gave Comet West little publicity.

Comet West (1976 VI). European astronomers were making a photographic survey of stars in the southern sky in 1976 when they noticed the image of Comet West (Fig. 2-2) on their photos. Comet West turned out to be everything that was expected of Kohoutek three years earlier. It was a big, bright comet with a breathtakingly beautiful tail and could be observed easily through binoculars from the Northern Hemisphere. Unfortunately the media still felt embarrassed over the anticlimax of Kohoutek's visit and, unwilling to risk another letdown, gave Comet West very little publicity. As a result this spectacular comet passed all but unnoticed by the public. Comet West was studied closely and gave astronomers a thrill by breaking up into four fragments as it rounded the sun. The breakup was heralded by a sudden flare from the head of the comet. At perihelion Comet West was roughly 20 million miles from the sun—considerably farther away than Kohoutek, which survived its solar encounter intact. Astronomers interpreted this evidence to mean that Comet West had a relatively "crumbly" nucleus. Probably the fragility of the nucleus was connected with the spectacular appearance of the coma (its bright head) and tail; the nucleus must have been "holed" in many places and thus highly productive of gas and dust. Some of the smaller nuclear fragments vanished completely under the fierce bombardment of solar heat.

THREE

The Rise of
Comet Science

EARLY CONCEPTIONS OF COMETS

One of the earliest mentions of comets in the scientific literature is a Babylonian text from around 1140 B.C. that describes a comet with a "luminous body" that shone "brightly as the day" and "a tail . . . like that of a scorpion." Some Babylonian astronomers believed that comets were fiery disturbances in the atmosphere, while some of their colleagues in the scientific priesthood held much the same view that modern astronomers do: namely, that comets are merely other celestial bodies like our earth, circling the sun as our planet does.

A few centuries later, as classical Greek civilization reached its peak, philosophers still disagreed about the nature and motion of comets. Scholars of the Pythagorean school, in the sixth century B.C., subscribed to the notion that comets were planets that did not keep a rigid schedule. We will see more of the Pythagoreans later, for their ideas, including the notion of a perfect heaven suspended above and around a corrupt and imperfect earth, had a great and largely detrimental influence on later astronomical thought.

Around 450 B.C., Hippocrates and his pupil Aeschylus proposed that while comets might be planetary bodies, their tails were merely optical illusions caused by reflected sunlight. (A similar optical-illusion hypothesis would be put forward some two thousand years later by no less an astronomer than Galileo.) Apollonius of Myndus, however, preferred the idea that comets were planets rather than mere tricks played on the eye.

The brilliance of comets was so like that of stars that many Greek thinkers saw comets as stellar objects. Two of Hippocrates' contemporaries, Anaxagoras and Democritus, suggested that comets were formed when drifting stars collided with one another and left a "splash" of light across the heavens. Democritus also suspected that comets might be the residue left behind when a star burned out or dissolved: the ashes, so to speak, of an extinct celestial fire.

Not all the ancient Greeks discussed comets in a calm and dignified spirit of inquiry. Sometimes the debate descended into calumny and accusations of fraud. After Euphorus of Cyme wrote that a comet in 371 B.C. had been seen to split apart, as comets sometimes do, Seneca replied with nasty slurs on Euphorus' character and charged him with making up the story to entertain the public. (Such low blows are not unknown among astronomers even today. When one United States astrophysicist, at a conference some years ago, presented solid evidence to suggest that Albert Einstein's famous general theory of relativity might be flawed, a well-known astronomer stood up in the rear of the hall and started shouting obscenities at him.) But when not engaged in slander, Seneca wrote some highly perceptive observations about comets. He thought them to be something very close to the modern view: a population of celestial objects with unusual orbits that carried them into view at unpredictable times.

Aristotle's "Atmospheric" Comets

The most influential of all the natural philosophers of classical Greece was Aristotle, who in the third century B.C. set down his thoughts on comets in his famous *Meteorology*. Aristotle reasoned that comets could not be planets because

all the planets were confined to the zodiac, the belt of constellations familiar to readers of modern horoscope columns
(Aries, Taurus, Gemini, etc.), and comets might appear in
extrazodiacal spots. He also rejected the idea that comets
were caused by conjunctions of planets or by colliding stars.

Aristotle thought they were phenomena of the upper air,
and his reasoning seemed sound at the time. Comets, he
pointed out, appeared at highly irregular intervals. Their
sightings were as unpredictable as the weather. Therefore,
he posited, they must be weather of some kind.

Aristotle ascribed the appearances of comets to warm, dry
"exhalations" of air from the lower atmosphere that caught
fire at higher altitudes and burned at just the right temperature to produce a lengthy and bright combustion, like a
well-adjusted gas flame in a modern home furnace. The
shape of the "burn," said Aristotle, determined the appearance of the comet. If burning took place at equal rates on all
sides, the result was a "fringed" comet without a prominent
tail, whereas a tail (or "beard") signified that the combustion was lopsided.

That was a convincing explanation in view of what was
known at the time about the heavenly bodies and the processes at work in the atmosphere. Eventually Aristotle and
his followers fitted comets into a comprehensive model of
the universe; and since that model shaped scientific thinking for hundreds of years afterward, we ought to look at it in
detail.

The Aristotelian universe was geocentric, meaning that
it had the earth at its center (Fig. 3-1). This also seemed a
fair assumption, for nothing in our everyday experience
suggests the earth we live on is moving through space.
Since we are traveling through space at the same velocity as
the earth, we have no sense of motion relative to the planet
and naturally have the impression that it and we are standing still while the rest of the universe—the sun, moon, other
planets, and stars—revolves around us.

In Aristotle's cosmology, the celestial objects were all
mounted on rigid crystalline spheres that circled the central

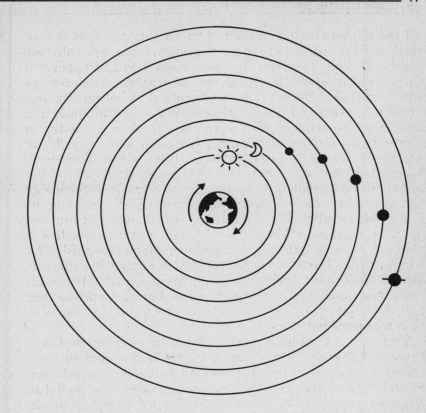

Figure 3-1. The Aristotelian System. In Aristotle's cosmology, the earth stood at the center of the universe and was surrounded by a series of transparent, rigid crystalline spheres on which were mounted the sun, moon, stars, and planets. Comets were believed to occur in the upper atmosphere. Though inaccurate, the Aristotelian model of the universe dominated astronomical thought for centuries.

earth and were turned by some kind of heavenly clockwork. The innermost sphere was occupied by the moon, while the outermost was the sphere of the fixed stars, an unchanging "shell" for the whole universe. Comets, according to Aristotle, occurred in the atmosphere between the earth and the moon.

Today, of course, this geocentric cosmology seems hopelessly archaic. We are used to thinking in terms of a suncentered (heliocentric) system and therefore tend to look down our noses at Aristotle. For its time, however, the Aristotelian model of the universe was a brilliant piece of scientific thought. It fitted all the known facts of astronomy and worked perfectly for most practical purposes. (Indeed, it still does work for many uses; our present-day system of celestial navigation, used by ships at sea, is based on Aristotle's ancient assumption that the earth is the center of the universe.) Only as more and more detailed observations of the heavenly bodies piled up, over later centuries, did scholars see that Aristotle's work was due for revision.

Aristotle's disciples made some minor changes to his grand scheme over the next few hundred years. Posidonius, in the second century B.C., thought comets were more numerous than Aristotle had supposed, because the glare of the sun hid them from view. Posidonius reached that conclusion after noticing a comet close to the sun during a total solar eclipse.

The naturalist Pliny the Elder, a credulous collector of tall tales, mentioned comets in his *Natural History,* published around A.D. 77, but added little to the work of Aristotle and Seneca. Pliny's main contribution was a comet classification scheme based on shapes and colors. Pliny shared the popular superstitions about comets being omens of things to come, as one passage from his writings shows: "We have, in the war between Caesar and Pompey, an example of the terrible events that follow the appearance of a comet. Toward the outbreak of this war, the darkest nights were illuminated . . . by unknown stars; the heavens seemed to be on fire; burning torches traversed in all directions the depths of space; the comet, that terrible star that overthrows the [governments] of the earth, showed its awful tresses." All the phenomena Pliny describes—"unknown stars," "burning torches," etc.—may be explained either as comets or as other phenomena associated with them, such as meteor showers. (We will examine the link between comets and meteor showers in a later chapter.) Shortly after Pliny published his classification system for comets, he was killed in the eruption of Vesuvius that buried the cities of Pompeii and Herculaneum.

The Ptolemaic System

Between Pliny's death and the second century A.D., comets began to take on an astrological significance. Contrary to widespread belief, there is a difference between astrological thinking and mere superstition; rather than simply linking comets in a generalized way to disasters, astrologers believed they could use comets' appearances to predict upcoming events well in advance, much as modern meteorologists use weather maps to predict rainfall and temperature for a given city several days in the future. This "astrologizing" of comets was largely the work of the second-century Egyptian astrologer Claudius Ptolemaeus, who is better known as Ptolemy but was no relation to the Egyptian royal family of that name. An impressive-looking man with a prominent nose and a luxuriant black beard, Ptolemy synthesized and condensed the astrological lore of his time into the comprehensive system that is used to draw up horoscopes today. When modern astrologers talk about the various houses, cusps, ascendant, descendant, and so forth, they are using the system that Ptolemy worked out for them more than eighteen hundred years ago.

Ptolemy considered comets to be only minor features of the heavens. He seldom mentioned comets in his writings, except to suggest that they might be used to forecast the weather. The influence of his work in astrology, however, encouraged the public to see comets as omens of events to come, and that attitude would help to surround comets with all manner of fears and superstitions.

After Ptolemy, comet science stood still for more than a thousand years. Comets were observed and their appearances recorded, but scholars (in Europe at least) seemed disinclined to analyze them as the ancient Greeks did. The

Europeans were satisfied to simply classify them, in the Aristotelian mode, as fiery storms in the upper air.

During the Middle Ages, the Christian church elevated Aristotle's work almost to the status of holy writ. The reasons for glorifying Aristotle were many. Among them were honest respect for his brilliance and his achievements. The church, however, had more self-serving reasons for paying the ancient Greek homage. Aristotle's vision of a geocentric cosmos tied in neatly with Christian ideas about the structure of the universe. Under the sway of Pythagorean thinking, early Christians conceived of heaven as the eternal, changeless, perfect abode of God, where everything happened in accordance with divine universal law. This concept of an ideal, unchanging heaven—"from age to age the same," as the revivalist hymn puts it—dovetailed nicely with Aristotle's vision of a sphere of fixed, unchanging stars at the edge of Creation. Aristotle's heavenly sphere model also appealed to the church thinkers, for they supposed (again under the influence of the Pythagoreans, who were forever reading metaphysics into simple geometry) that the sphere and circle were the most "divine" shapes. Since every point on a circle's rim or a sphere's surface is exactly the same distance from the center, those forms were thought to come as close as possible to the perfection of the Godhead.

By the dawn of the European Renaissance, progressive thinkers could see that the cosmology of Aristotle and Ptolemy would have to be either overhauled or scrapped entirely. The questions were: What would take its place, and who would go beyond the sainted Aristotle?

TYCHO BRAHE AND
THE COPERNICAN REVOLUTION

The man who started undoing the ancient earth-centered model of the cosmos was a Polish monk named Nicolaus Copernicus. He saw that the old cosmology was needlessly complicated, and began slashing away at it with a tool of logic known as Occam's Razor. Named after the Greek philosopher who first stated it, Occam's Razor is the principle that the simplest explanation of something is the most likely to be true.

Copernicus found a much simpler way to explain the observed motions of heavenly bodies. He put the sun at the center of the solar system and all the planets in circular orbits about it (Fig. 3-2). The Copernican scheme was concise and elegant, and it matched well with the observed motions of the sun, moon, and planets; but Copernicus's ideas came so close to heresy that he was forced to delay publication of his theory until after his death. Even then, the book wound up on the church's notorious Index of forbidden writings.

Fortunately for science, the Protestant Reformation loosened the Catholic church's grip on scientific thought and thus encouraged a new spirit of inquiry that would lead eventually to the vindication of Copernicus and make possible a revolution in astronomy. Among the leaders in that revolution were two men who spent long hours in the study of comets: the Austrian mystic Johannes Kepler and his mentor, the flamboyant Danish astronomer Tycho Brahe.

Tycho was, to put it mildly, a character. He was built like a powder keg, and the keg had an abnormally short fuse; the slightest hint of an insult would send him into a rage, and he had a genius for making enemies and starting quarrels. One of those quarrels disfigured him for life. Tycho had the bridge of his nose sliced off in a duel with a fellow student over which of them was the better mathematician. Tycho

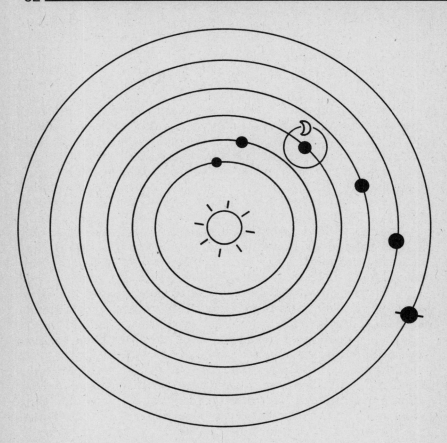

Figure 3-2. The Copernican System. Copernicus simplified the model of the cosmos greatly by assuming the sun to be the center of the solar system and the planets in orbit around it, with the stars far beyond the system's edge. He imagined the orbits of the planets were perfect circles, much as the Aristotelians had assumed the planets to be set in ideal spheres.

had an artificial bridge made out of gold and silver, and kept it in place with adhesive.

Tycho was born in what is now southern Sweden (then Danish territory) in 1546, and as a child witnessed a partial eclipse of the sun. The heavenly spectacle fascinated Tycho, and he decided to become an astronomer. It is probably more accurate to say that he was Denmark's official astrologer. His brand of astrology was not the kind seen in newspaper columns today ("Leo: This is a good time for travel"), but rather a much more sophisticated practice, which involved casting complex natal charts showing the positions of the planets at the time of an individual's birth. The natal chart supposedly showed what kinds of potential the subject had instilled in him or her by the planets. By looking at a chart Tycho believed he could foretell, in a general way, what manner of fortune and opportunities that person was likely to have. For example, Saturn (the planet that supposedly governed one's line of work) near the midheaven (the "top") in a natal chart indicated that the individual would probably be an outstanding success in his or her career, assuming that other aspects of the chart were favorable. The king, the nation, and the government all had their horoscopes, and Tycho's job was to peer into the future and see what might lie ahead for Denmark and its rulers. Tycho was well paid for his services and took his astrological duties seriously. "To deny astrology," he once roared at a skeptic, "is to deny the glory of God!"

Tycho's success as an astrologer and astronomer was due partly to the gratitude of King Frederick of Denmark. Tycho's father, a minor nobleman, had once saved the king's life during a military campaign and sacrificed his own life in the process. Frederick showed his gratitude to the Brahe family by putting Tycho on the government payroll and giving him an island called Hveen (Venus) for his residence. Tycho suddenly found himself blessed with a handsome income on top of the rents from his newly acquired lands. With that money Tycho built himself a mansion and an observatory that he named Uraniborg, after the Muse of astronomy. Next Tycho outfitted the observatory with the most sophisticated instruments that money would buy and set about keeping meticulous records of everything he saw in the heavens. His marvelous instruments and observations, as well as his painstaking approach to measurement, marked him as the first truly modern astronomer, and his observational data were a whole order of magnitude more accurate than any ever made before.

Tycho and the Comet of 1577

Among the things Tycho watched was a very bright comet that appeared in the autumn of 1577. In Tycho's day, the most widely accepted explanation of comets was still Aristotle's hypothesis that they were luminous storms in the atmosphere high above the ground. The Copernicans, however, were raising serious doubts about the accuracy of the Aristotelian system, which was no longer very good at predicting when and where celestial objects would appear. Sometimes a planet would show up days before or after the Aristotelian system indicated it should. What good was Aristotle's model, the Copernicans asked, if it was useless for making predictions?

Such talk made little impression on Tycho. He was a staunch defender of the old system of cosmology and to the end of his life did his best to preserve and protect the traditional model of the cosmos—though, as we will see in a moment, he was forced to modify that model slightly to make it more acceptable to the scientists of his day. Ironically, Tycho's own observations of a nova that appeared in the constellation Cassiopeia in 1572 did much to undermine the Aristotelian model by showing that the supposedly changeless "sphere" of stars was changing after all. Even more destructive to the earth-centered system, however, were Tycho's observations of the comet of 1577. Those data led inescapably to the conclusion that Aristotle and Ptolemy had been wrong.

Each night Tycho gathered as much information about the comet as he could. He noted the exact position of the comet's head and how far away it stood from several nearby fixed stars. He was trying to determine its distance from the earth. If the comet were near the earth (that is, within the hypothetical sphere of the moon) then the comet could be classified as an atmospheric phenomenon. If, on the other hand, the comet were farther away than the moon, then it would have to be assigned to the realm of the heavenly spheres.

Tycho's method involved parallax. This astronomical term denotes the difference in apparent position of a celestial object when sighted at two different locations. If you want to figure the parallax of a planet or star, first you must note its apparent position as seen from Point A on earth. Then take another sighting at Point B and compare the two. The difference between the two apparent positions, and the separation between Points A and B, will let you calculate the distance of the object.

Preferably, the two points ought to be as widely separated as possible. That way the change in the object's apparent position will be greater, and its distance can be figured more accurately. The larger the parallax, the nearer the object, because celestial bodies at close range shift their apparent positions more markedly than faraway objects do. The moon, for example, shows greater parallax than the planet Pluto, at the outer fringe of our solar system. When measuring the parallax of stars and galaxies, astronomers take their sightings six months apart, so that they will have a "base line," or distance between points of observation, some 185 million miles long.

Tycho didn't have to go quite that far. He measured the position of the comet's head among the stars when it was low in the sky and again when it was higher. Tycho was using the rotation of the earth to give him a good long base line of several thousand miles. All he had to do was wait a few hours between sightings.

He knew that if the comet were sublunar, or inside the moon's "sphere," he would come up with a parallax of one degree or greater. His parallax value for the comet, however, was only about 15 minutes, or one-fourth of a degree. This relatively tiny value showed that the comet was several times farther away than the moon—somewhere in the vicinity of 600,000 miles distant, compared to only about 250,000 miles between the moon and earth.

There was some disagreement with Tycho's finding. Hagecius, who enjoyed a prestige almost equal to Tycho's, carried out his own position measurements of the comet and achieved a parallax figure of 5 degrees, which would have put the comet in the sublunar realm. Unfortunately, Hagecius's instruments were much less accurate than those available to Tycho and threw off his calculations. Hagecius soon realized he was in error, and several years after the comet's passage admitted that Tycho had been right all along.

Not everyone, however, acknowledged Tycho's triumph. Much later, in 1621, a wily and unscrupulous astronomer named Claramontius attacked Tycho's measurements and calculations of the comet's distance in his book *Antitycho* and tried to prove that the comet really was sublunar. *Antitycho* was a worthless tract because Claramontius picked out only the observations that suited his thesis and ignored any data that indicated a superlunar position for the comet. About the same time, Tycho's findings were confirmed by the Italian Jesuit astronomer Oratio Grassi, a member of the prestigious Collegium Romanum, which was roughly equivalent to the modern-day National Academy of Sciences in the United States or the Royal Society in Great Britain. Grassi published a book in which he replicated Tycho's observations and came up with the same conclusion: the comet was indeed farther away than the moon. Once again Tycho came out on top.

Ironically, Tycho must have rued his own success, for his results only furthered the task that Copernicus began: the destruction of the Aristotelian system. It was difficult to accept Aristotle's scheme of the cosmos now that one of his "atmospheric" comets had turned out to be a distant celestial object in the same realm as the other planets. Moreover, Tycho's estimate of the comet's dimensions showed it was far larger than any disturbance ever recorded in the atmosphere. The tail was 22 degrees of arc in length, meaning it stretched out for millions of miles. Tycho also worked out an estimated orbit for the comet and found it was not a circle, but a flattened closed curve, "somewhat oblong, like [an] oval," in Tycho's words. Such orbits had no place in the system of rigid spheres imagined by the Aristotelians. If the crystal spheres really existed, then the comet's passage would have shattered them and sent their shards raining down on the earth, along with the suddenly unsupported sun, moon, stars, and planets.

Tycho's observations of the comet of 1577 did shatter the earth-centered view of the cosmos, and a less stubborn individual would probably have admitted that the old system had been wrong and the new Copernican system was correct. But not Tycho. He was too proud and too set in his ways to admit that he had ever been wrong about anything, and he refused to concede defeat. Instead, he attempted a compromise between the old and the new. He came up with a peculiar hybrid of the two systems, now known as the Tychonic system, in which the earth circled the sun, but everything else in the solar system revolved around the earth. This weird-looking arrangement was unworkable, but as we will see in a moment, it turned out to have considerable influence in Italy—much to the chagrin of another famous astronomer.

Tycho Meets Kepler

Tycho might have gone on to make even greater discoveries at Uraniborg had he not been his own worst enemy. He was anything but a benevolent landlord on Hveen; Uraniborg had a basement dungeon in which Tycho imprisoned peasants who fell behind on their rents or otherwise displeased him, and his tyrannical rule prompted his tenants to file a lengthy list of grievances against him with the government. At the same time, Tycho's prickly pride and quick temper made him enemies at the court of King Frederick, and only the king's good will prevented Tycho from getting sacked. Shortly after King Frederick died in 1588, his young successor, Christian IV, told Tycho to either mind his manners or get lost. Tycho resented being ordered about by a mere boy and decided to leave Denmark. He packed his belongings and moved with his family and servants to Prague, where he flattered the Holy Roman Emperor Rudolf II into giving him a post as imperial mathematician. There Tycho also met a young astrologer named Johannes Kepler, whose studies would help to resolve many of the mysteries that still surrounded comets and the universe in general.

Kepler was Tycho's opposite in almost every way. To his dying day, Tycho stood for the old-time cosmology, while Kepler was convinced Copernicus had the right idea. Tycho was stout, loud, gregarious, and fond of high living, and never tired of telling the world how brilliant he was. Kepler was slender, quiet, shy, and an ascetic tormented by a deep-seated sense of sin. Kepler's Christian faith brought him little comfort. His morbid religiosity verged on madness, and Tycho's worldliness and debauchery filled Kepler with horror. (Tycho was a champion boozer and died of a prostate problem caused by his excessive drinking.)

Tycho, for his part, at first disliked and distrusted this holier-than-thou young man, but eventually his attitude softened, and just before his death in 1601 the old Dane handed over to Kepler his priceless collection of observations. Kepler used Tycho's data to work out his three famous laws of planetary motion, which provided the mathematical foundation for Newton's law of universal gravitation. Kepler's laws are important to understanding why comets behave the way they do, so let us look at them briefly.

Kepler's first law made a correction in Copernicus's model of the solar system. Copernicus, remember, had assumed the planets' orbits to be perfectly circular. Tycho's data showed Kepler that the planets' orbital motions are better described by ellipses rather than by ideal circles (Fig. 3-3). Since this and Kepler's other two laws apply to comets as well as planets, we will substitute "comet" for "planet" in the rest of this discussion of Kepler's laws.

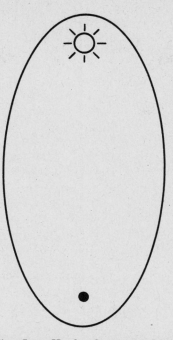

Figure 3-3. Kepler's First Law. Kepler determined that the orbits of planets were not ideal circles, as Copernicus had figured, but rather were ellipses—that is, slightly flattened circles—with the sun at one of their foci, or focal points.

Kepler's second law dealt with the velocity of objects in orbit. A comet travels more slowly as it moves farther away from the sun. Each day it sweeps out a wedgelike section, or "slice," within its orbit. Kepler measured the areas swept out by an orbiting comet and proved that the comet would sweep out equal areas in equal times. When the comet is close to the sun its velocity increases, and it sweeps out an area of X million square miles per day. As it later moves away from the sun, the "slice" then gets longer but narrower, and the comet slows down so that the area swept out daily remains constant.

Kepler's third, or "harmonic," law established a precise relationship between the diameter of a comet's orbit and the time (or "period") the comet takes to go once around the sun. The square of the comet's period—that is, the period multiplied by itself—is proportional to the cube of its average distance from the sun (distance × distance × distance).

So if you know a comet's average distance from the sun, you can use Kepler's third law to work out the comet's period, or vice versa.

Let's illustrate this law with a familiar example: the orbit of Halley's Comet. That comet has a period (P) of about seventy-five years. The square of P is 75×75, or 5625. So far, so good. Now, what about the average distance (d)? To simplify the calculation, we can measure the distance in terms of "astronomical units." One astronomical unit (a.u.) is the average distance between the earth and sun, or roughly 93 million miles. The cube of the distance is $d \times d \times d$. Now we can write a simple equation:

$$5625 = d \times d \times d$$

From here on things are easy. Just find a number the cube of which equals 5625. The answer is approximately 17.78. Therefore we know that Halley's Comet averages a distance of about $17\frac{3}{4}$ astronomical units from the sun, or somewhere in the vicinity of 1.65 billion miles.

Kepler's three laws look simple, and they are. Deriving them took him years of hard work, however (these were the days before adding machines and electronic calculators, and an astronomer might need months to complete a calculation that takes only a fraction of a second today), and Kepler could never have gotten the right answers without Tycho's supremely accurate data to work with. So it is easy to see why Kepler felt a debt of gratitude toward Tycho and sprang to the defense of his work and reputation on more than one occasion.

Kepler took on Claramontius shortly after *Antitycho* was published and easily countered his attack on Tycho. In his 1625 book entitled *Tychonis Brahei Dani Hyperapistes* (freely translated, *In Defense of Tycho Brahe the Dane*), Kepler proved beyond reasonable doubt the accuracy of Tycho's observations and calculations, and in so doing disposed of Claramontius's arguments once and for all. Kepler even defended Tycho's work successfully in the face of criticisms brought by Galileo.

Galileo's Mistake

Galileo is revered today as one of the secular saints of science, largely for his heroic stand against the tyranny of the church. Though he had used the newly invented telescope to prove the accuracy of the Copernican system, an ecclesiastical court made him recant on threat of torture and had Galileo placed under house arrest for the remainder of his life. Shameful and tragic as this incident was, it is hard to feel complete sympathy for Galileo, for in his years of freedom he snubbed or ignored many other scientists, including Kepler, who asked for his help and might have aided him in turn during his hour of need. Galileo also alienated many of his colleagues by his sharp tongue and acidic sense of humor; he was a master of sarcasm, which sometimes helped him make a point but won him very few friends.

Galileo's principal works about comets are his *Discourse on Comets* and his treatise *The Assayer,* studies of three bright comets that appeared in 1618, just prior to the start of the Thirty Years' War. The comets came along at a critical juncture in Galileo's career. Just then he felt betrayed by the Jesuits. Up to that time he had counted the Jesuits among his allies in the struggle to overthrow the old Aristotelian model of the cosmos, for the Jesuits were among the most progressive orders in the church and had great respect for the scientific method. The Jesuits had joined Galileo in advocating the Copernican system because it described the observed motions of the heavenly bodies much better than the old system did; but under pressure from the Holy Office, which was something like an ecclesiastical secret police, the

Jesuits backed off from their support of Copernicus and embraced the Tychonic system instead, for it seemed to represent the most acceptable compromise between the Copernican and Aristotelian systems.

Oratio Grassi wrote a paper in support of Tycho's model of the universe. When Galileo read Grassi's work, he reacted with fury. He scribbled curses in the margins of his copy of Grassi's essay and composed the *Discourse on Comets* as a retort. Galileo tried to convince his readers that comets were not material objects, as Tycho had figured them to be, but merely optical illusions produced by vapors that rose from the earth and caught the sun's rays upon reaching a high altitude in the atmosphere. Galileo claimed that no solid body could behave as the Tychonic system required, for Tycho, like Copernicus, had placed the sun, moon, and planets in perfectly circular orbits, all moving in the same direction; and in such a system, Galileo argued, there was no place for comets, which moved in unpredictable patterns and—by Tycho's own admission— appeared to follow noncircular paths through space. Galileo also pointed out how comets seemed to pop out of the sky fully formed, with tails pointing out behind them. That, Galileo said, was hardly the behavior one would expect of actual planetlike bodies, which wax and wane gradually in size and brightness as their orbits bring them near the earth and then carry them away. Consequently Galileo proposed that comets had no material existence at all, but were merely tricks played on the eye by the optical properties of the atmosphere.

Galileo's erroneous conclusions were due in part to his use of secondhand information. Galileo had not even observed the comets of 1618 himself, for he had severe arthritis and a hernia, which required him to wear a bulky truss that restricted his movements. These problems kept him away from his telescope while the comets were in sight, so he had to make do with the notes of other astronomers who had neither Tycho's superb instruments nor the Dane's genius as an observer. Indeed, Galileo did much of the research for his works on comets while lying in bed, listening to reports brought to him by friends who had watched them.

To protect himself from retribution by the church, Galileo published the *Discourse on Comets* under the name of one of his students. Galileo's distinctive style of writing made it obvious who the real author was, however, and Grassi directed a counterblast at Galileo in his next book, *The Astronomical and Philosophical Balance.* In this case "balance" meant the kind of scales used to give fair weight at a market. That image conveyed an impression of justice and equitability and gave Grassi the air of a serious scholar simply trying to find out what was right. The image was also misleading, for Grassi was not bent on uncovering the truth, but only on hitting back at Galileo. Since Galileo had attacked him under cover of his student's name, Grassi invented the anagrammatical pen name "Lothario Sarsi" for his *Balance.* "Sarsi" came directly to the point. He demanded to know if Galileo was challenging the authority of the church by condemning the Aristotelian and Tychonic systems.

Grassi might have done better to publish the book under his own name, for the pseudonymous nature of "Sarsi" gave Galileo a splendid opportunity. The church could not legally prosecute him for attacking a fictitious person, so he was free to lambaste "Sarsi" with every rhetorical weapon at his command. The result was *The Assayer* (in Italian, *Il Saggiatore*). Galileo chose the title shrewdly. The "balance" in the title of Grassi's book was a crude device accurate to perhaps an ounce or so, but the "assayer" mentioned by Galileo was a person who measured gold dust and other precious materials as precisely as possible, down to the weight of the proverbial mustard seed. So Galileo's title alone put Grassi on the defensive, for it cast "Sarsi" in the role of a bungler whose work was full of inaccuracies. (Grassi, however, also knew how to play this game. He referred to *Il Saggiatore* as the *Assagiatore,* or "wine-sipper," thus implying that Galileo had been drunk when he wrote it.) *The Assayer* brimmed over with sarcasm, directed both at

Grassi and at Tycho. Galileo had special venom reserved for Tycho's comet studies. He referred to the comets as Tycho's "monkey-planets," or figments of the imagination.

To Kepler, Galileo's thinking was nonsense, and in an appendix to his 1625 defense of Tycho he disposed of Galileo's criticisms handily while outlining his own views on the nature of comets. Kepler suggested that a comet consisted of a ball of matter that had part of its substance "driven away" by the impact of sunlight, and thus "breathed out" its own tail. Basically, Kepler had the right idea. His vision differs only in a few details from the model of comets that is accepted by most astronomers today.

The Orbit Question

Kepler had less success in working out the comets' orbits. Though he had shown that the planets follow elliptical orbits rather than circular ones, Kepler never applied the ellipse equations to the orbits of comets. Instead he imagined that comets moved like skyrockets along straight lines (Fig. 3-4) and eventually burned themselves out as rockets do.

This straight-line model is not as farfetched as it may sound, for many comets travel in orbits so elongated that they look like pairs of parallel straight lines running through the solar system. Yet the inward and outward legs of the orbits do come together at perihelion and aphelion (the point farthest from the sun) to form a closed curve. The closing of this curve at perihelion is plainly visible to comet watchers on earth, as the comet traces out a smoothly curving path like a plunging roller coaster.

Why did Kepler assume comets traveled in straight lines, when the curvature of their orbits near perihelion was so clear to see? Kepler, in a curious parallel with Galileo, assumed that his eyes were playing tricks on him. The comets' orbits merely appeared to be curved, said Kepler, because the earth was traveling in a curved orbit that twisted the perceptions of human observers slightly.

The straight-line model of cometary motion bothered

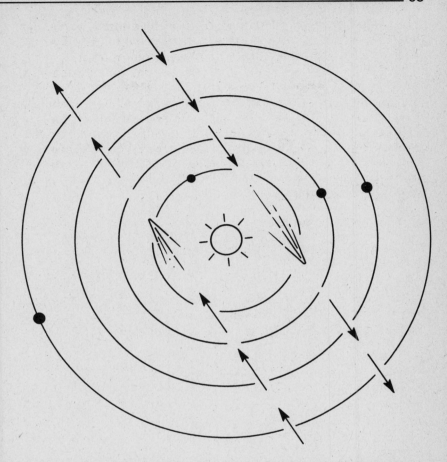

Figure 3-4. Kepler's Mistake. Kepler assumed that comets were not orbiting the sun, but rather, were shooting through the solar system in straight lines, like skyrockets.

some astronomers, among them Johann Hevelius of Danzig, who in his 1688 book *Cometographia* proposed that the comets' orbits did not merely appear to be curved, as Kepler had it, but really were very slightly curved. That sounded

plausible—more so, at any rate, than Hevelius's other beliefs about comets. He suggested that comets were not big
balls of matter, as Kepler had imagined, but rather were
disc-shaped—like giant poker chips—and were ejected in
great planetary "burps" from Jupiter and Saturn. The newborn comets spiraled outward from those worlds for a while
in a manner reminiscent of baby birds testing their wings,
and then sped away toward the sun. Hevelius's cometary
discs were supposed to keep one face pointed directly toward
the sun at all times, through a rather improbable process in-
volving a resistant interstellar medium that made them
twirl (Hevelius imagined) like flipped coins.

About the time *Cometographia* appeared, Giovanni
Borelli, writing under a pen name to protect himself from
Galileo's fate, came up with a better idea. He proposed that
cometary orbits were actually parabolas, or long open
curves, with the sun nestled in the bend of each curve. Borelli thought comets swooped in from deep space, made a
quick loop around the sun, and then returned to the dark
abyss. It was a beautifully simple and workable solution to
the mystery of comets' orbits, for it fitted in with the Copernican model of the solar system and accounted for the markedly curved paths of the comets near perihelion. The curve
of their orbits was real, said Borelli, and not an illusion.

The astronomers were getting close to the true picture of
how comets travel around the solar system, but the comple-
tion of that picture had to wait for the work of two gifted
Englishmen, Isaac Newton and his friend and colleague
Edmond Halley. Halley played such an important role in
Newton's career that the two scientists are often mentioned
together in the same breath as "Newtonandhalley," the astronomical equivalent of Gilbert and Sullivan.

FOUR

Newton, Halley, and the ——— Comet of 1680

HALLEY'S COMET AND THE *PRINCIPIA*

An avid stargazer from his earliest years, Halley wanted to contribute to the mapping of the skies, but in Europe he faced stiff competition from better-known astronomers such as Cassini and Hevelius. So Halley decided to study the stars of the Southern Hemisphere, of which there were few good observations. The island of St. Helena seemed to him the best place to scan the southern skies, and after obtaining government backing for his project, Halley set out for the island in the autumn of 1676. The India Company, which controlled the island, cooperated with Halley's research, but the weather did not. Clouds and mist obscured the night skies for much of his visit, and in the end he was able to make observations of only about 350 stars. Nonetheless, he had his work published in London in 1679, following his return to England.

Halley's star catalogue showed how quickly the Aristotelian system had crumbled and how thoroughly the Copernican system was established in Western astronomy by the late seventeenth century. Halley noted that several stars in the constellation Sagittarius were much dimmer than they were described in the writings of Ptolemy. "This appears to show," Halley wrote, "if not corruptibility, at least the changeability of the celestial bodies." Halley also demonstrated that some stars had changed their positions since Aristotle's day—another sign of an evolving universe. His star maps earned him the nickname of "the Southern Tycho."

Edmond Halley carried out many other projects in astronomy but is best remembered for his research on the subject of comets. He became interested in them while on a trip to the Continent.

In 1680 Halley, then twenty-four years old, was visiting Paris when a very bright comet appeared in the sky. He was awed by the spectacle and went to the Paris Observatory to obtain detailed records of the comet's movements from the observatory's director, Giovanni Cassini (whose principal claim to fame is his discovery of the gap—known as Cassini's division—between the ring systems of Saturn). Halley tried to plot the comet's path on the basis of that data but was frustrated because he made the same mistake that Kepler had made years earlier: He assumed the comet was moving in a straight line. Cassini suggested that the comet might be traveling on an orbit that kept it very near the sun, but this pattern also failed to match the observations. Finally, Halley was forced to put his calculations aside and return to his comfortable home in the London suburbs.

In 1682, Halley studied another comet from his home observatory. Watching the big, bright comet swinging across the sky, Halley wondered what laws of nature governed its progress. Kepler's laws of planetary motion described the movements of celestial bodies well enough, but in a sense they only scratched the surface; they were merely expressions of some underlying, more fundamental, law or laws of nature that would explain why the universe hangs together as it does. What, Halley thought, kept the stars, planets, and comets in their courses? Did a universal principle of gravitation bind them all together like cosmic "glue," and thus unify the solar system and presumably the rest of the cosmos as well?

A few decades earlier, the French philosopher René

Descartes, inventor of analytic geometry, speculated that gravitation was the result of huge "vortices," or cosmic whirlpools, swirling around in the universe. He thought the currents of space converged here and there into colossal vortices, within one of which the planets of our solar system whirled around the sun like ships caught in a maelstrom. Descartes had trouble working comets into his vortex model, for the comets did not spin around in neat, nearly circular orbits like those of the planets, but rather, sliced across the planets' orbits at unpredictable times and places.

At last Descartes came up with what seemed a convincing explanation. He suggested that comets were starlike objects that somehow had slipped out of the vortices where they formed and were weaving their way from one vortex to another. That explanation sounded good in theory but failed to account for comets with "retrograde orbits," meaning they traveled in the opposite direction to the observed motion of the planets. If a giant vortex was sweeping the planets along, then how could a comet swim against the current, so to speak, and achieve a retrograde orbit? This was only one shortcoming of the Cartesian vortex theory. It also was deficient in mathematical rigor. What scientists needed was a solid law of universal gravitation: something into which you could plug numbers and get a quantitative result. The lack of a law of gravitation had plagued astronomers for centuries and sometimes forced them to erroneous conclusions. For example, Kepler proposed that the moon's influence on the waters of the earth causes the tide to rise and fall. Today we know that Kepler was correct, but Galileo decided that Kepler was mistaken, simply because Galileo had no knowledge of a law of gravitation that would link the moon with the tides.

Halley was eager to find such a law but lacked the necessary training. So the following year he approached the physicist and chemist Robert Hooke, then president of the Royal Society, and discussed with him the idea of a universal law of gravitation. Hooke was intrigued by Halley's ideas, and the famous architect Sir Christopher Wren was persuaded to put up a modest cash award to be given the first person to work out the appropriate equations. No one Halley contacted had the mathematical skill for the job, however, until he went to see Newton at Cambridge.

By this time Newton was already a legend. He was also slightly mad, partly because of his intense shyness and distrustful nature, and partly because his research in alchemy was slowly poisoning him with heavy metals such as arsenic and mercury. That poisoning resulted in strange behavioral disorders, including symptoms of paranoia. Newton appears to have thought the whole world was out to get him, and this fancy probably explains why he was reluctant to publish his work and thus bring himself to public attention. Nonetheless, Newton was friendly toward Halley, who had a charming manner and was able to put the touchy Newton at ease. When Halley put the problem of universal gravitation to him, Newton revealed that he had worked out that law earlier but had misplaced his calculations.

Halley's jaw must have fallen. Newton had deciphered the riddle of gravitation, the goal of countless astronomers and mathematicians for more than a century, and had never mentioned it to anyone before. Halley asked him if he could reproduce the calculations. Newton said yes and promised to send them to him. Some months later Halley had Newton's figures in hand. A new era in science was about to begin.

Halley knew there was a gold mine of scientific discovery in Newton's uncollected writings. If only he could be talked into writing all his work down, science would be immeasurably richer. So Halley cajoled and badgered Newton into writing a book. The result was his monumental *Philosophiae Naturalis Principia Mathematica (Mathematical Principles of Natural Philosophy)*, commonly known as the *Principia*, in which he outlined the mathematical basis of physics and astronomy. Much of the *Principia* has to do with comets, and so one might say that comets helped to launch the Newtonian revolution in science and mathematics

—a revolution that changed the entire course of world history.

The *Principia* might never have seen the light of day without Halley's prompting and financial backing. Halley was by no means wealthy, yet he personally helped underwrite the publication of Newton's work, though Newton was reasonably well-to-do and could have met the publication costs himself. Halley even read the proofs of Newton's book for him as they came back from the printer. Greater devotion hath no friend.

The *Principia* was some of the heaviest reading in history, for Newton wrote it in Latin and did not stoop to simplify his math for the convenience of readers less gifted than he; but the *Principia* was also one of the most influential scientific treatises ever published, and it divided the history of physics neatly into pre-Newton and post-Newton epochs.

"I [will] now demonstrate the frame of the system of the world," Newton wrote in the *Principia*. Demonstrate it he did. He went far beyond Kepler and Tycho and formulated a set of general principles and equations from which Kepler's laws of planetary motion could be derived. Kepler had been dealing with special cases: the small picture, one might say. Newton saw instead the big picture, and his vision was so breathtakingly accurate that it dominated the physical sciences for many generations to come. Newton viewed the universe in terms of absolute time and absolute space and imagined a system in which the various parts of the cosmos were all linked to one another like gears and cogs in a single great machine. The Newtonian view of the universe prevailed for more than two hundred years and was modified only when Albert Einstein conceived his special and general theories of relativity in the early twentieth century.

The heart of Newton's system was his universal law of gravitation. Like many revolutionary ideas, it was basically simple. Newton said all objects in the universe exert a mutual attractive influence on one another. This influence is called gravitation, or gravity. Its effect is inversely proportional to the square of the distance between any two bodies. So if you double the distance between two objects, you reduce their gravitational pull on each other to one-fourth of what it was previously. Tripling the distance results in one-ninth the former pull. The equation may be written as follows:

$$G = \frac{1}{D^2}$$

Here G stands for the gravitational effect, and D represents the distance between the two bodies involved. This epoch-making breakthrough in science looked dull in print, however, and Newton and Halley decided to use a spectacular example to illustrate the universal law of gravity: the comet of 1680.

Borelli had suggested years earlier that a comet's path around the sun was a parabola, an elongated U-shaped curve with the sun nestled in the crook of the U. Newton agreed. A keen observer, he saw a comet moving away from the sun, on the sun's opposite side, shortly after the comet of 1680 was seen to disappear in the sun's glare. It seemed almost sure that the two comets were actually one and the same, swinging about the sun on a parabolic path; and since a parabola is nothing but the "tail end" of an extremely stretched-out ellipse, Newton had every reason to assume that the 1680 comet was traveling on an exceedingly long elliptical orbit, and therefore a member of the solar system. As Newton explained in the *Principia:*

[The comet's orbit] is determined . . . in an ellipse. And it is shown that [the comet moved] with as much accuracy as the planets move in the elliptic orbits given in astronomy.

Newton argued that comets were planetlike objects that circled the sun "in very eccentric orbits." He wrote, "I am out in my judgment if [comets] are not a sort of planets revolving in orbits returning into themselves with a perpetual

motion. . . . the comets range over all parts of the heavens
. . . they pass easily through the orbs of the planets, and
with great rapidity. . . ."

Galileo's influence was tremendous, however, and some
astronomers still agreed with him that comets were little or
nothing more than optical illusions produced by the refraction, or bending, of the sun's rays. Newton quickly disposed
of this notion. The mere fact that we can see the tails, he argued, indicates that they have a material existence just as
the earth does, for we could not see them at all unless they
contained matter that reflected sunlight to our eyes. Newton also pointed out that if comet tails were made up solely
of refracted sunlight, as some astronomers argued, then
comets would show colors (the familiar rainbow effect) like
those produced by prisms and lenses. Since comets showed
no such colors, Newton concluded that they did not shine by
refraction but rather by reflected sunlight, and therefore
had to be material.

Nonetheless, some scientists still clung to the view that
comets could not be solid, material objects because they
changed appearance quickly and therefore must be made of
insubstantial stuff, like the clouds. For those scientists,
Newton had a reply. He proposed that what observers saw
was not the solid body of the comet itself, but only the comet's surrounding envelope of vapor. The heads of comets
Newton thought to be "encompassed with huge atmospheres, and the lowermost parts of these atmospheres must
be the densest; and therefore it is in the clouds only, not in
the bodies of the comets themselves, that these changes are
seen. Thus the earth, if it was viewed from the planets,
would, without all doubt, shine by the light of its clouds, and
the solid body would scarcely appear through the surrounding clouds. . . . much more must the bodies of comets be
hid under their atmospheres, which are both deeper and
thicker." Later Newton added, "the bodies of the comets are
solid, compact, fixed, and durable, like the bodies of the
planets; for if they were nothing else but the vapors or exhalations of the earth, of the sun, and other planets, [a] comet,

in its passage by the neighborhood of the sun, would [be] immediately dissipated. . . ."

Newton also thought he saw a parallel between the small
masses of comets and the distribution of planetary masses
in the solar system: little planets close to the sun, giant
planets farther out. Newton wrote that "as in the planets
which are without tails, those are commonly less which are
revolved in lesser orbits, and nearer to the sun, so in comets
it is probable that those which in their perihelion approach
to the sun are generally of less magnitude, that they may
not agitate the sun too much by their attractions." Newton
was overestimating both the masses and the gravitational
influence of comets. Not even the largest known comet has
enough gravitation to "agitate" the sun noticeably. Otherwise, however, he was very close to the picture of comets
that is accepted by most astronomers today.

Newton himself did not carry out the calculations needed
to prove his assertion that the comet were material,
planetlike objects following elliptical, planetlike orbits
around the sun. He merely suggested the comet circled
the sun in an elliptical path, and then left the long-
suffering Halley to calculate the orbit. Halley kept Newton informed of his progress by mail, and an excerpt from
one of Halley's letters to Newton provides an idea of what
Halley was up against:

. . . having done the comet of 1683, which I can represent
most exactly, and that of 1684 (wherein I find Hevelius
has not been able to observe with exactness requisite) . . .
I fell to consider that of 1680/81, [that is, the comet of
1680, which lingered into the early part of 1681], which
you have described [in the *Principia*], and looking over
your catalogue of the observed places, I find that in that
of the 25th of January 1681, there is a mistake of 20 minutes in the longitude of that day . . . I find certain indication of an elliptic orb in that comet and am satisfied that
it will be very difficult to hit it exactly by a parabolic.

Halley had good reason to be concerned about the accuracy of his data, because elliptical and parabolic orbits (or "orbs," as Halley called them) are almost identical in close proximity to the sun, and so the slightest error could spell the difference between getting the orbit right and missing it altogether. After much labor, Halley got it right. As Newton had assumed, the comet followed an elliptical path around the sun, and the comet's path could be described precisely by Newton's equations.

Thus Newton and Halley proved the validity of the universal law of gravitation and dispelled much of the mystery that still surrounded comets. No longer was there anything mystical or supernatural about them. Newton's math and Halley's observations showed that comets were, like the earth and the other planets, nothing more than material objects in space, subject to the same natural laws as any other celestial bodies.

In addition, Newton made some shrewd surmises about the physical processes at work on comets. Because comets were most often seen close to the sun, he concluded that they "shine by the sun's light, which they reflect." If comets shone by their own light, he pointed out, then they would be seen equally often in the sun's vicinity and in the distant reaches of the solar system; "yet, looking over the history of comets, I find that four or five times more have been seen in the hemisphere towards the sun than in the opposite hemisphere." He also believed that a comet's tail was part of its "atmosphere," and that the atmosphere was generated by the action of solar heat upon the solid body of the comet. Discussing one comet, Newton pointed out that "in the month of December, just after it had been heated by the sun, [the comet] did emit a much longer tail, and much more splendid, than in the month of November before, when it had not yet arrived at its perihelion; and, universally, the greatest and most fulgent tails always arise from comets immediately after their passing by the neighborhood of the sun. Therefore the heat received by the comet conduces to the greatness of the tail: from which, I think, I may infer that the tail is nothing else but a very fine vapour, which the head or nucleus of the comet emits by its heat." The comet's apparent capacity for heat astounded Newton. A comet must soak up "an immense heat from the sun and [retain] that heat for an exceedingly long time" to keep the tail gases spewing out, Newton wrote in the *Principia*.

Newton would not have found the process so amazing had he known that comets are made up largely of frozen water and carbon dioxide, and that not too much heat is needed to turn those ices into vapor. (We will examine the postulated structure of comets in detail in a later chapter.) Newton suspected—correctly, as it turned out—that the total amount of matter in the tail was very small, despite the tremendous dimensions of the tail. "But that the atmospheres of comets may furnish a supply of vapour great enough to fill so immense spaces, we may easily understand from the rarity of our own air; for the air near the surface of our earth possesses a space 850 times greater than water of the same weight; and therefore a cylinder of air 850 feet high is of equal weight with a cylinder of water of the same breadth, and but one foot high." What happened to the material in the tail after the comet passed by? Newton thought the tail stuff continued in a "state of rarefaction and distillation" until it was "at last dissipated and scattered through the whole heavens, and little by little . . . attracted toward the planets by its gravity, and mixed with their atmosphere" to provide a "spirit" vital to life. "I suspect," wrote Newton, "that it is chiefly from the comets that spirit comes, which is indeed the smallest but the most subtle and useful part of our air, and so much required to sustain the life of all things with us." This idea is not as quaint as it may sound. Modern scientists are finally coming to realize what Newton recognized centuries ago: Comets do appear to have a major influence on life on our planet, though not quite in the "spiritual" way Newton imagined. Keep that idea in mind, for we will return to it later.

At one point Newton speculated that inflammable comets falling into a star might produce a nova. He even suggested that the nova observed by Tycho Brahe in 1577 was caused by an infall of comets. "So fixed stars," wrote Newton, "that have been gradually wasted by the light and vapours emitted from them for a long time, may be recruited"—that is, refueled—"by comets that fall upon them; and from this fresh supply of new fuel those old stars, acquiring new splendor, may pass for new stars."

Here Newton was wrong. We know today that a nova results when a star is exhausting its "fuel," not when it is receiving more. Newton was correct, however, to assume that some comets fall directly into the sun and are destroyed. The death of such a "kamikaze" comet was photographed in 1981 by a U.S. Naval Observatory camera experiment on a Defense Department satellite. The experiment, called SOL-WIND, was designed to photograph the corona, or atmosphere, of the sun by blocking out the solar disc and thus bringing the corona into clear view. On January 26, SOL-WIND photographed a small comet approaching the sun. The comet then vanished in the glare of the sun and never reappeared on the other side. The comet had either plowed directly into the sun or passed so close that it was annihilated by the fierce solar heat. One of the cosmic events that Newton envisioned three centuries earlier had finally been captured on film.

At the end of his hardheaded analysis of comets and their place in the universe, Newton allowed himself some theological ramblings. "This most beautiful system of the sun, planets, and comets, could only proceed from the counsel and dominion of an intelligent and powerful being . . . and on account of his dominion he is wont to be called Lord God over all. . . ." Newton goes on in this vein for another several pages, wondering about the nature of God and how God might animate the universe. Passages like this would be grossly out of place in modern scientific treatises, but in Newton's day science and religion were still strongly intertwined, and many serious scientific works were as much theology as science. Later we will see how one of Newton's colleagues at Cambridge used hypothetical comets to account for some of the alleged miracles mentioned in the Old Testament, and in so doing inaugurated a tradition of crackpot literature that has endured to this day.

The *Principia* was the crowning glory of Newton's career. He never attained that height of brilliance again, for he suffered a nervous collapse around 1690 and was too deranged to do serious scientific work thereafter. Apparently the long hours spent in his alchemy lab took their toll, for Newton, as Jonathan Swift put it, aged like a tree: he began to die first at the top. Newton was able to work for the government even in his impaired state, however, and served as master of the mint and as a member of Parliament. Newton's biographer Eric Temple Bell points out that Newton's contributions to the deliberations of Parliament were minor; apparently his only spoken words there were a request to open a window. In spare moments Newton devoted himself to theological studies and the unraveling of biblical chronology. Like Einstein after him, Newton ended his days as little more than a figurehead—a totem set up for public veneration.

While Newton slipped into his decline, Halley was still thinking about comets. The riddle of their orbits had been solved. Now what about their periods? If comets circled the sun in elliptical orbits, Halley reasoned, then their return trips could very likely be calculated and predicted with a great degree of reliability. He thought he would find such a pattern in the recorded appearances of comets over the centuries.

Halley reviewed the literature on comets and gathered every reliable sighting available to him. He found very few before the fourteenth century. Important data were often missing from earlier records. There was the bright comet of A.D. 1066 that supposedly doomed King Harold, but its path was not carefully recorded. Nicephoras Gregoras had left an accurate report of the path of a comet sighted in 1337, but he did not record the exact dates of its passage. Only in the fifteenth century did Halley begin to find observations precise and detailed enough for his purposes. Regiomontanus made detailed observations of the great comet of 1472, and

in the following century Tycho started keeping his exqui-
sitely accurate records of comets. Kepler, Hevelius, and
Borelli contributed still more sightings for Halley's research.

Finally, Halley had a desk full of data and began to look
for recurrent patterns in them. He knew that any comet in a
reasonably stable orbit would keep showing up, time and
again, on a regular schedule. If any of those comets had distinctive features such as unusual size and brightness, they
ought to stand out clearly from the rest, both by their visual
characteristics and by their regular reappearances.

One comet in particular stood out from the rest: the comet
of 1682. Halley discovered that a comet matching its description had appeared in 1607, 1531, 1456, 1380, and so on,
back to A.D. 1066. This comet—assuming it really was the
same comet appearing over and over again—had a period of
seventy-five to seventy-six years. Halley checked his figures
for the orbit of the comet of 1682 and found that its estimated period was the same.

Halley felt confident that he had found a periodic comet.
Yet some little irregularities in its period bothered him. If
the comet really were governed by the Newtonian laws of
gravity and of motion, then shouldn't it be coming around
as regularly as clockwork? Shouldn't the comet take exactly
so many days, not one day more nor less, to circle the sun
and reappear in the skies of the earth?

In theory, yes; but in practice, the comet was slightly less
than punctual. One of its orbits might be a few months
longer or shorter than the previous one. Did this "play" in
the comet's orbit mean that Newton had been wrong, or
might something out there be swaying the comet slightly
out of its previous orbit and off its expected schedule?

Halley thought the culprits to be the giant planets Jupiter and Saturn (Fig. 4-1). In a letter to Newton he wrote, "I
must entreat you to consider how far a comet's motion may
be disturbed by . . . Saturn and Jupiter, particularly in its
ascent from the sun [that is, its return to deep space], and
what difference they may cause in the time of the revolution

Figure 4-1. Saturn. The ringed gas giant has a powerful gravitational field that can change the orbits of comets and delay their returns by days or even weeks.

of a comet in its so very elliptic orb." As it happened, Halley was right again. The terrific gravitational fields of Jupiter and Saturn were more than sufficient to yank a relatively tiny body like a comet in one direction or another. The giant planets' effects notwithstanding, Halley felt confident enough about his findings to stick his neck out and predict when the comet of 1682 would reappear. "I . . . predict its return in the year 1758," Halley wrote. He also boasted that if he turned out to be right, then the world would know that the period-icity of comets "was first discovered by an Englishman."

The French, however, were first to recalculate the orbit of the comet of 1682 allowing for the gravitational influence of Jupiter and Saturn. A team of mathematicians under the direction of the astronomer Alexis Clairault, using only pencil and paper, computed the comet's orbit and predicted that it would reach perihelion in mid-April of 1759. The French team was off by only four weeks. The comet was sighted on Christmas Day, 1758, and attained perihelion on March 13, 1759. The comet of 1682 had shown up again as Halley had predicted, and henceforth would be known as Halley's Comet. Halley did not live to see his prediction come true, for he died in 1742 at the ripe old age of eighty-five.

Together, Newton and Halley accomplished a revolution in science. They put celestial mechanics, the physics of planetary motion, on a sound mathematical basis while stripping comets of the last vestige of their supernatural aura and proving them to be mere physical objects under the influence of physical laws. At the same time, Halley and Newton demolished the Cartesian scheme of vortices, by showing (in Newton's words) how "comets . . . move every day with the greatest freedom, and preserve their motions for an exceeding long time, even when contrary to the course of the planets."

Halley's study of comets also helped bring order to the often chaotic historical records of their appearances. He demonstrated that many individual comet sightings were really repeated appearances of periodic comets that swung around the sun once every few years or decades.

Astronomers have identified more than two dozen returns of Halley's Comet from historical records, starting in 239 B.C. On that trip the comet is said to have inspired the Carthaginian general Hamilcar to make his son Hannibal swear everlasting enmity toward Rome. On later visits Halley's Comet was implicated in the deaths of Agrippa (11 B.C.) and the Emperor Macrinus (A.D. 218), the Mongol invasion of China and India (A.D. 1222), a Dutch dike failure in which four hundred thousand persons were drowned (A.D. 1531), and a war between Native Americans and English settlers in Virginia (A.D. 1607). That last event gave rise to a famous bit of Americana. King Powhatan and his warriors saw the comet as a "red knife in the sky" and thought it portended a good chance to drive the white invaders from their lands. So Powhatan attacked the settlers and took Captain John Smith prisoner. Smith was about to be executed, the story goes, when Powhatan's daughter Pocahontas intervened to save his life. Modern accounts of this romantic tale seldom mention that Halley's Comet started it all.

SOME FAMOUS COMET HUNTERS

Comet watching became a mania among astronomers in the century after the 1758 return of Halley's Comet, and the French carried on the tradition of Clairault's team by taking the lead in searching out comets. The greatest comet hunter in Europe, and perhaps of all time, was Jean-Louis Pons. He was working as a porter at the Marseilles Observatory when, in 1801, he discovered his first comet. The director of the observatory was so impressed by Pons's observational skills that he made him an assistant astronomer. Pons went on to become Astronomer Royal at Lucca, and he came to be known as the "ferret of comets" for spotting more than thirty others in his career.

Other highly successful comet hunters in France included Pierre Méchain, Montaigne de Limoges, and Charles Messier, a clerk at the Paris Observatory. Messier used his knowledge of comets to compose propaganda for the regime of Emperor Napoleon I, and once wrote a pamphlet entitled "The Marvelous Comet that Appeared at the Birth of Napoleon the Great," linking Napoleon's nativity with great events in the heavenly sphere. (Messier is better remembered for a catalogue of faintly luminous celestial objects that looked like comets but were not. He made note of these objects so that other comet watchers would not mistake

them for comets. More than a century later, astronomers
with telescopes much more powerful than Messier's would
focus on Messier's "noncomets" and discover them to be distant galaxies. That is why many galaxies are designated
with the letter M—for Messier—followed by a number: the
French comet hunter discovered them without knowing
what they really were.) Keeping ahead of his rival Montaigne became an obsession with Messier. When his wife
came down with a terminal illness shortly after he discovered his twelfth comet, Messier stayed away from his telescope to care for her, and in the meantime Montaigne
discovered a comet that Messier thought was rightfully his.
Later one of Messier's colleagues heard of the death of Mme.
Messier and expressed sorrow at the astronomer's loss.
"Alas, Montaigne has robbed me of my thirteenth comet!"
Messier exclaimed. Then he realized the reference was to
his wife rather than the comet, and he added: "Oh . . . poor
woman."

Perhaps the most famous American comet hunter was
Edward E. Barnard, who lived from 1857 to 1923 and was
probably the only astronomer who ever used comets to pay
off his home mortgage. He was interested in astronomy as a
young man, and in 1881, shortly after his marriage, he won
a two-hundred-dollar prize for discovering a comet. Barnard
used that money as a down payment on a house, and hoped
that by spotting more comets he could earn enough addi-
tional prize money to pay off the mortgage. He wrote later
that he and his wife "could look forward only with dread to
the meeting of the notes that must come due. However, the
hand of Providence seemed to hover over our heads; for
when the first note came due a faint comet was discovered. . . ." More and more comets swam into Barnard's field
of view over the years, and soon he was able to pay off the
bank and boast that his house "was built entirely of comets."

FIVE

———— Comet Science
———— Comes of Age

THE CALCULATING PROBLEM

Newton and Halley made it possible for astronomers of the late seventeenth and the eighteenth century to calculate the orbits of comets and forecast their returns with previously unthinkable accuracy. This victory seemed Pyrrhic at times, however, because of the sheer magnitude of the calculations required. Whole teams of mathematicians needed entire years to calculate a comet's path around the sun, allowing for factors such as the gravitation of the giant planets Jupiter and Saturn. In those days before the advent of the computer and desktop calculator, a single problem in celestial mechanics could fill reams of paper and consume a considerable part of an astronomer's life. Moreover, there was always the danger that errors would creep in through fatigue or carelessness; and even a mistake of a single digit could make the results of the calculation useless. Often astronomers were well along in a calculation when they discovered an arithmetical error earlier in their work and had to start over again. Kepler said he had to carry out one complicated calculation seventy times before he finally got it right. (Kepler came agonizingly close to getting his hands on a calculating machine that would have made his job much easier. His friend Wilhelm Schickard, a mathematician and inventor, devised a mechanical computer that could handle numbers up to 999,999 and could carry out addition, subtraction, multiplication, and division. Schickard was about to send Kepler one of his calculators when a fire destroyed the workshop where the machines were being assembled, and the one intended for Kepler went up in smoke. Before Schickard could put together another calculator for him, Schickard died of pestilence spread by troop movements during the Thirty Years' War. Had Kepler had access to Schickard's invention, his research in astronomy might have progressed much faster.)

The pain of astronomical calculations in the late eighteenth century is conveyed by the words of a French astronomer named Lepaute, who helped calculate the date for the return of Halley's Comet in 1759:

> For six months we calculated from morning till night, sometimes even at meals; [as a result] I came down with an illness that affected my constitution for the rest of my life. [Without] the help provided by Madame Lepaute . . . we should never have dared to undertake the enormous labor that was necessary to calculate the distance of each of the two planets, Jupiter and Saturn, from the comet, separately for every degree, for 150 years.

Partly because of the terrific burden of the calculating process, more than a century passed after Halley's prediction before anyone else successfully forecast the return of a comet. Many scientists tried to simplify the calculating procedure but had only limited success until the early 1800's.

About the time Napoleon I rose to power, the French mathematician Pierre Simon de Laplace proposed a method of "successive approximations" that looked good in theory but was clumsy and time-consuming, so Laplace did little to relieve the burden of calculation. (Laplace knew Napoleon personally and is famous for one of his remarks to the little emperor. Napoleon, who was a comet enthusiast and is said

to have interpreted comets as omens of future success in battle, also fancied himself a great mathematician and once
spent a few hours reading a treatise on celestial mechanics
that Laplace had written. Later he asked Laplace why his
work made no mention of God. "Sir," Laplace replied
haughtily, "I have no need of that hypothesis!")

Then an amateur astronomer named Wilhelm Olbers succeeded where Laplace failed. Olbers, who made his living as
a physician, created a quick and accurate method for calculating a parabolic orbit for a comet. Unfortunately, not all
comets had parabolic orbits. Some were clearly elliptical,
and for them another method was needed.

Enter Karl Friedrich Gauss. A bricklayer's son, he is
widely regarded, with Newton and Einstein, as one of the
three greatest mathematicians of all time. He rose from
poverty to become the master of Germany's world-renowned
scientific research center at Göttingen, which contained one
of Europe's most famous observatories and made many important contributions to nineteenth-century astronomy.
Gauss himself dabbled in astronomy from time to time, and
around the turn of the century he became involved in a
search for a "missing planet" between the orbits of Mars
and Jupiter. Kepler's mathematical model of the solar system indicated that a planet ought to be found in that zone,
but no planet was visible. Astronomers were convinced that
something had to be there, however, and so a worldwide
search for the missing world began.

Anything sighted in between Mars and Jupiter would
need to have its orbit calculated. If the object had an orbit
that kept it more or less in between the orbits of Jupiter
and Mars, then astronomers would know they had finally
plugged the hole, so to speak, in the solar system. To aid in
this quest, Gauss devised a mathematical technique that
could be used to calculate an elliptical orbit quickly and easily, from only three sightings. Here at last were the tools
that comet watchers had needed, and the work of Gauss and
Olbers bore fruit almost immediately.

Encke's Periodic Comet

Only a few years later, Johann Encke was studying observations of several comets that appeared in 1786, 1795, 1805, and 1818, and he thought he saw a regular pattern in the sightings.

These comets were all assumed to be one-time visitors to our skies, and to have parabolic orbits, signifying that they zipped in once from deep space and then returned to the dark void forever. Yet when Encke looked at the orbits calculated for these comets, he could see the orbits were not parabolas, but ellipses! Elliptical orbits meant that these comets would return; and in 1819 Encke discovered that all these observations were sightings of the same comet, coming back time and again at an interval of roughly three and a half years. Its perihelion was roughly 40 million miles from the sun, and its aphelion some 375 million miles away—approximately four times the distance between the sun and earth.

Encke used Gauss's mathematics to work out the comet's orbit, and Encke's calculations indicated that the comet would return in 1822. The comet did so and is now named after Encke, though he himself always called it the Comet of Pons in honor of the famous comet hunter who was among the first to observe it. (The comet was actually discovered by Méchain in 1786.) Encke's Comet, officially known as P/Encke, is still visible, spinning around the sun approximately once every forty months.

Encke's discovery, however, uncovered another mystery about comets, a mystery that for a time cast doubt on the validity of Newton's work. The comet was obeying Newton's laws of motion, but not very rigorously. Newtonian theory said the comet ought to be keeping a definite schedule, right down to the day, hour, and minute; but in this case theory seemed to fall slightly short of practice, because Encke's Comet did not show up exactly when expected. Instead, it ran consistently ahead of time. On each return, the comet's period was found to have decreased by two to three hours. Something was moving the comet into a tighter and tighter orbit, cutting an average of eight seconds per day off its orbital period. Those seconds and hours could not be explained by the gravitational effects of Jupiter and Saturn, nor by any other known influence in the solar system.

Well, so what? Newton's physics accounted for more than 99.9 percent of the comet's orbital period. The leftover minutes amounted to less than one-tenth of one percent of the comet's "lap time" around the sun. Wasn't that degree of accuracy good enough?

Unfortunately, not for astronomers. The discrepancy looked small but was still too big to ignore, for where astronomical distances are involved, even the tiniest percentage error in an estimate of orbital period can translate into a difference of millions of miles between the expected and observed positions of a planet or comet. This gap between prediction and observation put Newton's work in an unfavorable light. After all, the test of any theory is its predictive capability, and inaccurate predictions means a theory is probably wrong. Even a single mistaken forecast, imprecise by the tiniest fraction of one percent, may bring a theory tumbling down. Had Newton therefore been in error? Were there exceptions to his seemingly all-embracing scheme of the universe, or was some unknown (but still Newtonian) influence out there acting on Encke's Comet to give it the mere appearance of flouting Newton's laws?

To make things even more puzzling, the comet—almost as if it realized itself to be unpunctual—began making more regular returns soon after Encke revealed its periodic nature. The discrepancy shrank to two hours . . . then one hour . . . then only a few minutes, until today the comet's period diminishes on each orbit by about the duration of a commercial break on television. Did the comet carry some kind of "stopwatch" to correct errors in its Newtonian timetable? These questions bedeviled astronomers for most of the nineteenth century and well into the twentieth.

The erratic behavior of Encke's Comet was only one of the

enigmas that surrounded comets in the 1800's and early 1900's. The makeup of the tail was another riddle for many years after Encke's discovery. Of what substance was the tail composed, and what made it stream away from the sun? No astronomers still thought seriously, as Galileo did, that the tail was simply an optical illusion; Kepler and Newton had laid that misconception to rest. Yet the composition and dynamics of the tail remained largely unknown.

NEW INSTRUMENTS AND DISCOVERIES

A revolution in observational techniques, starting in the mid-1800's, helped to answer these and hundreds of other questions about comets. Before then, astronomers had to spend long hours pressing their eyes close to telescope lenses; and even with large telescopes to help, there was a limit to how much detail the human eye could make out. All too often it was difficult to tell whether some feature of the comet, such as a bright spot or curious swirl in the tail, really existed or whether it was a figment of the observer's imagination, like the "hideous human faces" and "blood-stained swords" that Paré fancied he saw in the comet of 1528.

Recording telescopic observations of comets posed another problem. Verbal descriptions and drawings could convey only a limited amount of information and were distressingly prone to errors and distortions. Students of such records frequently had to ask themselves: Did a drawing show an actual detail of the comet, or was that "detail" just a smudge on the paper?

Photography changed all that. As soon as cameras could be fitted to the eyepieces of telescopes, astronomers no longer needed to sit for hours and strain their vision to make out features of comets. The comet's image was captured on photographic plates quickly and accurately and could be studied later at leisure.

Also, the camera could do things that were impossible for the human eye. An astronomer's camera can gather light over a period of minutes, hours, or even days, and build up a "time exposure" that reveals details invisible to the unaided eye. The faintest features of a comet stand out boldly when captured in a time exposure. All astronomers have to do is keep the telescope pointed directly at the comet, using a clock-controlled motor to move the telescope barrel on its mount, for the duration of the exposure.

An Englishman named Usherwood was the first to photograph a comet successfully, in 1858. After an exposure of several seconds, he was rewarded with a recognizable picture of the comet, tail and all. A quarter of a century later, photography was a widely used tool for comet study. Plates exposed over periods of minutes or hours revealed many heretofore unsuspected details of comet structure, from small "eddies" and "knots" in the tails to beautiful helical structures that stretched from the tip of the tail all the way to the nucleus. Some comets appeared to be pirouetting across the skies, trailing streamers behind them. Other comets gave the impression of having "engine trouble," for they left in their paths ragged puffs and gouts of tail substance like the exhaust from a balky locomotive. There were comets with the faintest of tails, visible only after long hours of plate exposure. There were comets with tails so seemingly dense and solid that one imagined one could reach out and grab them like broom handles. There were comets with halos, comets with "horns" projecting sunward, comets with intricate traceries of luminosity about them like the fountains at Versailles.

Cameras were not the only instruments that could be attached to telescopes. Beginning in the 1860's, astronomers curious about the chemical composition of comets joined the telescope to the spectroscope, a device that uses a prism to break down the light from comets and other celestial objects into its component wavelengths, known altogether as its spectrum. The pattern of light and dark lines in the spectrum is called a spectrogram when recorded on film, and looks a bit like the "zebra stripe" product codes seen on many consumer goods today. Each "stripe" stands for light emitted or reflected or absorbed at a given wavelength. Every substance has a characteristic spectrogram, just as every product on a supermarket shelf has its own individual code. There are certain telltale lines unique to carbon monoxide, water vapor, and anything else that may be found in space. By mounting spectroscopes on their telescopes, astronomers could tell, even from millions of miles away, what comets were made up of. An Italian astronomer made the first spectroscopic observations of a comet in 1864 and saw several fuzzy "emission bands" that indicated the comet was giving off light of its own rather than merely reflecting light from the sun. There was then no way to tell what substances produced those lines, for spectroscopy had not yet established the characteristic spectral "fingerprints" for large numbers of chemicals. Advances followed fast, however, and in 1866, when British astronomers repeated their Italian colleague's observations on another comet, they saw a spectral pattern of reflected sunlight superimposed on the emission bands. That pattern indicated that comets were both reflecting and emitting light. It was interesting to note that comets combined features of nonluminous bodies like planets and luminous objects like stars.

Also remarkable (when astronomers finally learned to match the bands with the substances to which they belonged) was the similarity between the chemical composition of comets and meteorites. Experiments showed that meteorites, when heated, gave off gases similar to those observed in the tails of comets, including hydrocarbons such as ethylene (C_2H_4). Were meteorites, then, fragments of comets that had fallen to earth? That seemed a reasonable assumption, as an astronomer named Kirkwood wrote in 1861: "May not our periodic meteors be the debris of ancient but now disintegrated comets whose matter has become distributed round their orbits?" The Italian astronomer Giovanni Schiaparelli confirmed Kirkwood's suspicion when, in 1866, he discovered that meteor streams were indeed associated with the orbits of comets. Schiaparelli linked the famous Perseid and Leonid meteor showers respectively with the comets Swift-Tuttle and Tempel-Tuttle.

(To avoid confusing meteors with meteorites and meteoroids, we had better define them all before going further. A meteoroid is a natural object several miles or less in diameter drifting along through outer space. A meteorite is a meteoroid that has fallen through the atmosphere and landed on the earth, while a meteor is the stream of luminous gas, commonly called a shooting star, that a meteoroid produces on its fall through a planet's atmosphere. Only a small proportion of incoming meteoroids will reach the ground. Most are no bigger than grains of beach sand or aquarium gravel, and burn up completely on their drop through the atmosphere, leaving bright meteors to mark their passage. Meteors are precisely what Ptolemy and some of his Greek predecessors thought comets to be: luminous disturbances in the upper air.)

Spectroscopic studies of comets revealed much more than a chemical similarity to meteorites. Comet spectra also showed the presence of ionized nitrogen molecules (N_2+), ionized carbon monoxide $(CO+)$, cyanogen (CN), and many other components. "Ionized" means an atom or molecule has a positive or negative electrical charge that is generated when something—say, bombardment by high-energy radiation—produces an imbalance between the number of negatively charged particles, or electrons, and positively charged particles, or protons. A positive ion has a net surplus of protons, while a negative ion is "top-heavy" with electrons.

Positive ions such as $CO+$ and CO_2+ were prominent in comet spectra. When comets passed very close to the sun, the spectral bands of iron and other heavy heavy metals began to show up too. Each comet appeared to be a chemical cornucopia spewing out all manner of elements and compounds through its tail and into space.

Ionization (that is, the electrical "charging" of atoms and molecules) offered a possible explanation for the perplexing

behavior of comet tails. They stood out in a manner reminiscent of a flag in a strong wind, even though outer space was
known to be a vacuum (or so near one as to make no difference for most practical purposes) and therefore had no wind,
as we know it on earth, to waft the tails away. What provided the interplanetary "breeze" was a mystery.

To complicate matters, comets had not merely one but
several different kinds of tails (Fig. 5-1). Type I tails were
long and straight. These were the kind that Origen described as resembling beams of wood. Type II tails were
shorter and slightly curved (beardlike, in Origen's classification), while Type III tails were briefer still and showed a
pronounced curve like that of a sickle—or Origen's wine

Figure 5-1. Comet Tails. Comets display three kinds of tails. Type 1 tails
(left) are long and straight and consist almost entirely of hydrogen and
other gases. Type 2 tails (center) are more curved and contain more dust,
while Type 3 (right) are very rich in dust and show an even more pronounced curve.

containers, which in his day were curved and had broad bases and narrow spouts. Some comets displayed more than one type of tail at the same time. Donati's Comet, in 1859, trailed a wide, curved Type II tail and two long, narrow Type I tails that stuck out on one side of it like the antennae of a lobster.

Bessel, the same astronomer who erred in regard to the orbit of Biela's Comet, invoked electrical charges on tail material to explain why comets' tails behaved as they did. He had seen a comet expel an antitail toward the sun. The antitail stood out in front of the comet like the bowsprit of a ship. Then, as Bessel watched, the sunward-pointing spurt bent backward and joined the main mass of the tail. Some influence from the sun was repelling both the tail and the antitail. Bessel assumed that electricity must be the answer.

The sun, Bessel proposed, had a strong electrical charge that repelled the similarly charged matter in a comet's tail and forced it into a long linear configuration. (Oppositely charged particles show mutual attraction, while identically charged particles, whether positive or negative, will keep each other at a distance.) If Bessel had the right idea, the sun repelled a comet tail because the sun and the tail material both had the same kind of charge on them.

Bessel had come up with a clever idea of how comets got their tails, but his electrical-charge hypothesis did not explain the three different types of tail configurations—I, II, and III. If the tail's dynamics were simply a matter of the charge on the sun repelling the charge on the tail, then why didn't all comet tails look virtually identical instead of showing different degrees of curvature?

Other factors, astronomers decided, must also help to give comet tails their various shapes. Maybe chemistry could offer an answer. Spectrograms showed that the tail stuff was not a single homogeneous substance, but rather, a mixture of elements and compounds that reacted in different ways to the force (whatever it was) from the sun. So another comet expert named Bredichin modified Bessel's hypothesis in light of what was becoming known about the chemistry of comets. Bredichin believed their different chemical "recipes" determined what kind of tails the comets would display. He suggested that Type I tails were made up primarily of hydrogen, the lightest element known, which would be "blown away" easily—like smoke in a high wind—to form long, straight, narrow tails like the "antennae" displayed by Donati's Comet. Bredichin thought Type II tails were generated by hydrocarbons like ethylene, and Type III tails by heavy metallic vapors. The heavier the substances in it, the more the tail "drooped." Yet this was speculation, not fact, for the nature of the "tail-wagging" influence from the sun was still unknown; and not everyone was happy with the ideas of Bessel and Bredichin.

Sweden's famous chemist Svante Arrhenius was dissatisfied with the electrical explanation and proposed instead that sunlight produced the "gusts" that blew comets' tails around. Though sunlight seems insubstantial to us here on earth, the sun's rays exert a faint but steady pressure on objects out in space: not much of a force, but enough to overcome solar gravitation and push atoms, molecules, and tiny dust particles away from the sun. Physicists in Arrhenius's day calculated how strongly sunlight could nudge a bit of spaceborne dust, and they came up with a surprising answer. Under some conditions, the pressure of sunlight could exceed twenty times the sun's gravitation! So maybe sunshine, not electricity, was responsible for blowing the tails around.

Unfortunately, similar calculations showed that light pressure by itself was inadequate for the job. Light might exert enough pressure to produce a sluggish Type II or III tail, but the long, straight Type I tails were possessed of much greater velocities than light pressure alone could achieve. By Bredichin's day, observations of comet tails were so detailed and accurate that it was possible to calculate the speed of particles in the tail; and Type I velocities were awesome. Whatever was blowing Type I tails away, it accelerated the tail molecules to velocities of more than 150 miles per second in some cases, or approximately three hundred times faster than a bullet fired from a revolver. A particle traveling at that velocity, and meeting no resistance, could cross the Gulf of Mexico in the time it takes to dial a phone number.

The Solar Wind and Comets

Type I tail velocities required a "shove" two hundred to two thousand times stronger than solar gravitation: ten to one hundred times more than light pressure alone could supply. Even if one went beyond the wavelengths of visible light and added the pressure of high-energy (and to our eyes invisible) ultraviolet light, the "sun-tanning" rays at the far end of the spectrum, sunlight still had insufficient energy to blow tails around unaided. A more reasonable candidate for the interplanetary "breeze" was particle radiation —the so-called solar wind. Essentially, the solar wind is an endless stream of charged hydrogen atoms spewed out from the sun, and is quite different from sunlight, though the two kinds of solar radiation are sometimes confused with each other.

A light wave is an insubstantial thing. It exists somewhere in the borderland between matter and mathematics, and may be seen as slightly more than a mere wave function and slightly less than a particle like an electron. A particle in the solar wind, on the other hand, is a full-fledged chunk of matter that, when it hits something, packs a far greater punch than a wave of light does. If you were a gas molecule in a comet's tail, the pressure of sunlight might feel to you like a brisk March wind, whereas a particle's impact would be more like getting tackled by a three-hundred-pound linebacker.

Astronomers and physicists were aware of the solar wind as early as the late 1800's, when the sun was discovered to spew out charged particles, known then as corpuscles or corpuscular rays. Soon after their discovery, these particles were linked with the spectacular nighttime auroral displays in the polar regions, commonly called the northern and southern lights.

Around the turn of the century, two Norwegian scientists familiar with the aurora, physicist Olaf Birkeland (the scientist who trekked all the way to the northernmost tip of Scandinavia to observe the northern lights during the 1910 passage of Halley's Comet) and his mathematician colleague Carl Størmer, suggested that the aurora lit up at night when the magnetic field of the earth caught charged particles whooshing outward from the sun and funneled them down into the atmosphere around the earth's north and south magnetic poles. There the particles interacted somehow with air molecules to make them glow. Birkeland drew an analogy between the aurora and the luminous patterns seen on a cathode-ray tube (like the picture tube in modern television sets) when streams of charged particles struck the face of the tube from inside. Størmer tried to work out a theoretical explanation for the swirling patterns of light that the incoming particles made overhead, but he failed. Even today we are unsure how the sun's particle output generates the aurora, but Størmer and Birkeland get much of the credit for setting astronomers and physicists to thinking about the solar wind and its influence on the rest of the solar system.

Scientists soon realized that the solar wind also provided a plausible account of why solar flares—great eruptions on the surface of the sun—are followed in a day or two by powerful magnetic storms here on earth that play havoc with radio and phone communications. British geophysicist Sydney Chapman demonstrated in the 1930's that the particles in a solar flare shooting out from the sun at approximately a thousand miles per second (slightly more than 3.5 million miles per hour), would sweep past our world one or two days later and shake the earth's magnetic field, much as a strong wind tosses the leaves and branches of a tree. This unusually strong gust in the solar wind, said Chapman, gave rise to a magnetic storm.

The solar wind's effects were not confined to the earth,

either. American researchers studying cosmic rays, the extremely energetic radiation that bombards our world constantly from space, found that the cosmic ray count dropped off sharply when the solar wind was gusting. Particle radiation from the sun was sweeping cosmic rays out of the solar system like dust in a gale, as the charge on the ionized solar wind particles repelled the charge on the incoming cosmic radiation. If the solar wind could blow away cosmic rays and make the earth's whole magnetic field shudder, then it surely packed enough energy to sweep comet tails away from the sun. That is exactly what happens. The wind from the sun sweeps past comets and carries their "exhaled" gas and dust away toward deep space. So Bessel was partly right when he claimed that an electromagnetic effect pointed comet tails away from the sun, because the particles in the solar wind have a predominantly positive charge, as do the ionized gases in a comet's tail. Positives push away other positives, and consequently the tail gases are swept away in the breeze from the sun.

This breeze is highly tenuous. When it gets to the earth's orbit, the solar wind contains only some four thousand particles per cubic inch: about the same density as a pinch of pepper spread out through the air in Madison Square Garden. Yet that rarefied zephyr sweeping out from the sun is more than adequate to brush away a comet's tail, because the tail gases are almost equally thin. A thousand cubic miles of tail contain barely enough matter to fill a saucer, and British astronomer Sir John Herschel was exaggerating only slightly when he said that an entire comet could be "packed in a portmanteau," or suitcase.

The solar wind also explains the three different tail types. As Bredichin supposed, Type I tails are almost entirely gaseous, and the solar wind blows them away most easily, producing a long straight configuration. Types II and III tails are dustier; the relatively heavy particles in them put up more resistance and are slower to blow away, so that they tend to form scimitar-shaped tails like the one that dismayed Josephus when he saw it hanging in the heavens over Jerusalem. Bredichin, like Bessel, had been on the right track. He had merely assigned the role of the dust grains to heavy atoms and molecules.

Part of the tails' curvature results not from their composition, but rather from the pattern of the solar wind itself. The particles in it do not stream out from the sun in straight lines like the spokes of a wagon wheel. Instead, the sun's rotation puts a curve on the particle streams, and they spiral outward in a pattern like a whirlpool in reverse. This same pattern is imparted to the dust and gas in a comet's tail, so that when you see a comet curving across the sky, you are also seeing the solar wind made visible.

Although the solar wind theory accounted for the dynamics of the tail, the tail had to originate from somewhere, and therein lay still another mystery. To what was a comet's tail attached? Obviously something was "breathing out" the tail gases, as Kepler and Newton had surmised long before, and that something had to be a solid object or collection of such objects.

The problem was, the tail's source was so tiny compared to the rest of the comet that it was impossible to observe directly. A comet's solid core, or nucleus, compares in size to its coma roughly as a baseball in Candlestick Park does to San Francisco; and if you could magically shrink the coma to the size of San Francisco, then the comet's tail would reach all the way to Canada.

The nucleus was therefore so infinitesimal, and so thickly shrouded in view-obscuring vapors, that ordinary observational techniques were useless. Astronomers would have to deduce the nucleus's properties from observations of the entire comet, just as nuclear physicists were forced to deduce the properties of the atom's tiny nucleus by studies of whole atoms. There was no other option, unless there were some way to lay hands on samples of a comet nucleus and bring them into a lab for analysis.

Had any parts of a comet nucleus ever fallen to earth? Maybe they had (if Schiaparelli was correct), in the form of meteorites. So scientists looked to meteorites for clues to the character of comet nuclei.

A typical meteorite is a pocket-sized stone weighing about a pound. It has a dark surface, an interior much the same color as cigarette ash, and a low weight for its volume. Though commonly thought to be rare, meteorites are actually widespread on earth. The soil in an average suburban yard may contain several pounds of gravel-sized meteorites, and specimens weighing ten pounds or more are by no means uncommon. You may have a weathered meteorite or two, indistinguishable at first glance from ordinary stones, decorating your garden or your front walk.

Meteorites are so abundant, and were linked so conclusively with cometary orbits by Schiaparelli, that scientists around the turn of the century decided that a comet nucleus must be made up of a closely packed swarm of meteoroids, ranging in size from a small fraction of an inch to the diameter of boulders, and held together by mutual gravitational attraction. The individual meteoroids in the swarm supposedly gave off gases when heated by near approaches to the sun, just as meteorites did when heated in a laboratory. First the liberated gases would form an atmosphere of sorts —the coma—around the nucleus, then would be brushed away by solar radiation to form the comet's tail.

Sandbanks and Snowballs

This conception of comet nuclei came to be known as the flying sandbank model, and during the 1940's the renowned British mathematician R. A. Lyttleton and several of his colleagues came up with an elegant and intriguing explanation for the sandbank's origin.

Lyttleton's study of comets was a *tour de force* of mathematical analysis. His math is beyond the scope of this book, but in very general terms, this is what he said:

Occasionally the sun passes through a huge cloud of interstellar dust made up of material cast out of ancient novas in their death throes. These "ashes" of expired stars are initially only a few millionths of an inch in diameter. Solar gravitation draws the individual dust particles sunward so that they tend to converge on an imaginary line—the accretion axis, Lyttleton called it—in the sun's wake. The accretion axis runs through the sun parallel to the direction of the sun's forward movement. There the particles bump into one another and come to rest in small aggregations that get steadily bigger and eventually go into orbit around the sun as flying sandbanks. Mixed in with the dust are gas molecules that were also swept up along the accretion axis and packed in with the dust aggregations. This gas is released to form the comet's coma and tail on close approaches to the sun.

What happened in the Lyttleton model is comparable to a phenomenon that can sometimes be seen on dusty highways. Turbulence in the wake of a passing car stirs up dust particles. If the dust is allowed to settle undisturbed, much of it will come to rest along a line parallel to the direction of the car's movement, about midway between the tracks of the wheels. This line corresponds to the accretion axis in the Lyttleton model. Here the process involved is swirling air rather than solar gravitation, but the outcome is much the same. If these two situations, the one in space and the other on the road, were completely analogous, then the car would reach its destination surrounded by little orbiting swarms of fine grit like the flying sandbanks circling the sun in Lyttleton's scheme.

The flying sandbank model had some points in its favor. It explained reasonably well the connection between meteor streams and the orbits of comets, as well as the similarities between the chemicals in comet tails and in the gases trapped in meteorites. When most of the volatile gases had been "cooked out" of the meteoroids by solar heat, all that supposedly remained was a cloud of rocky "buckshot" that manifested itself in the form of meteor showers when the rocks encountered our atmosphere. Meteorites appeared to be the bigger bits of the sandbank that survived their fall through the atmosphere and landed on the earth's surface.

The sandbank model also agreed with the low mass estimates for comet nuclei. Everyone could see that comets were only tiny bits of matter by astronomical standards. No comet had ever been seen to sway a planet even slightly off course, and during the comet scare of 1857, Jacques Babinet came close to the correct proportions when he compared the masses of a comet and the earth respectively to those of a fly and an express train. Laplace figured that no comet ever observed had more than 1/5,000, or 0.02 percent, the mass of the earth. That was a liberal estimate, for a single foothill in the Appalachians weighs more than a typical comet nucleus does. A sandbank-centered comet would have almost no mass compared with the sun, earth, or other planets—a few thousand tons at most—and therefore the sandbank model was consistent with most known cometary mass values.

Yet there were problems with the sandbank model. For one thing, it was based to a large extent on conjecture instead of fact. It is one thing to assume a connection between meteorites and comet nuclei, and quite another thing to demonstrate it. There was (and still is) no proof of any known meteorite having originated in a comet nucleus. The "smoking gun" that would tie in comets to meteorites was missing.

This was only one of the relatively minor flaws in the flying sandbank model. It failed some more important tests as well. For instance, why do comets hold together? Obviously they do. Many are so durable that they have been undergoing a grueling road race around the sun every few years for centuries with no apparent ill effect. But it was hard to reconcile the ruggedness of comets with the flimsy structure imagined for them in the flying sandbank model, which indicated that comets ought to be fragile and extremely short-lived phenomena, not familiar friends that came back time and again.

To show how the sandbank model flunked this test, let's perform what the Germans call a *Gedankenexperiment,* or "thought experiment," in the imagination. Picture a tightly clustered swarm of meteoroids in an orbit that carries them near the sun. Far away from the sun, the particles hang close together because no nearby body has enough gravitation to pull them apart, unless they happen to pass too close to Jupiter or Saturn.

As the swarm approaches the sun, however, Kepler's laws of motion start to have a dramatic effect on the particles. The closer particles are to the sun, the faster they move in orbit. Now notice that each individual particle in our hypothetical cloud is traveling in its own orbit. Some particles are closer to the sun than others and therefore will move a little faster as they approach perihelion.

The result? Particles on the "sunny side" of the swarm will tend to outpace those on the "shady side" away from the sun, so that the swarm will stretch out and dissipate. A few turns around the sun would suffice to smear the sandbank all over the sky and destroy the tight concentration of particles that the model required.

So a sandbanklike nucleus couldn't hold together for very long. Yet many comets, including Encke's and Halley's, are long-lived. They have been seen to return frequently, with their nuclei as tightly packed as ever. That fact alone was strong evidence against the sandbank model.

Another objection had to do with gas jets that were observed coming from the nucleus. Photos showed that fountains of gas were shooting out from distinct locations on the nucleus like water from the nozzles of firehoses. This effect gave some comets multiple tails; Borelli's Comet, for example, sported nine tails on its visit in 1903. That pattern was hard to explain in terms of the sandbank model, for the particles in its nuclear swarm would release gases more or less evenly in all directions—somewhat in the manner of a melting snowman—rather than in concentrated streams.

The gas jets and the longevity of comets made two major strikes against the sandbank model. The third strike that put that model out, so to speak, came from estimates of the tail's total mass. According to the sandbank model, meteoroids in the nucleus gave off the gas that streamed away to form the tail. Rocks, however, could not release that much material. A typical comet, on a single turn around the sun, loses hundreds of times more matter than a swarm of meteoroids could provide. A flying sandbank could therefore sustain a visible tail for only one or two days at most. In no case could a sandbank-centered comet blaze in the sky for several days or weeks, let alone repeat the performance time and again over a span of centuries. The sandbank would literally run out of gas.

For that matter, how could a sandbank comet survive the sun's intense heat at all? A single pass near the sun would totally vaporize a sandbank comet, even if its component particles were made of asbestos.

One would therefore expect sandbank comets to be "cooked away" completely upon or prior to rounding the sun. Yet most comets have most of their substance left after perihelion, for they can be seen swinging around the sun and heading back into deep space. So they had to be considerably tougher than the sandbank model allowed.

These objections and others doomed the sandbank model. By 1950 many scientists were convinced that comets—as Newton surmised—were generated by cohesive, solid bodies of some kind instead of Lyttleton's meteoroid swarms, and astronomer Fred Whipple of Harvard University came up with a new model that accounted for virtually all the major features of comets. Whipple called his model the dirty snowball. He pictured the comet nucleus as a solid object maybe one or two miles wide, composed of dust, rock, and ice. As the comet nears the sun, solar heat starts making the ices sublimate into gas. The sublimation process also releases fine dust that was trapped in the icy matrix of the nucleus, and together the liberated gases and dust form the tail.

Whipple's dirty snowball model quickly supplanted the flying sandbank concept, for the same reason that the Copernican system had replaced the Ptolemaic one centuries earlier: the new model was simpler than the old one and explained observations much better. Whipple's model wiped out the discrepancy between the masses of comet tails and the gas supply of the nucleus. Even a tiny snowball nucleus, only a couple of thousand feet wide, could furnish more than enough gas to create a long and brilliant tail.

The Whipple model also explained why comets tend to remain intact rather than fall apart on their circuits of the sun. Unlike the unconsolidated sandbank of Lyttleton's model, the rigid nucleus Whipple envisioned could stand up to the stresses of a solar passage with no trouble. It could also round the sun hundreds of times without dissipating, since Whipple figured a typical comet loses only a fraction of one percent of its volatile components on each encounter with the sun. Indeed, the attrition rate might very well decrease as the ices in the comet's outer layers sublimated away, leaving behind an insulating "shell" of dust and grit that would protect underlying ices from the sun's destructive rays. This process would give the comet its own "sunscreen" and prolong its life.

The snowball nucleus was by no means indestructible. It could be fragmented, as was Biela's Comet. The fragments of a shattered snowball nucleus would continue through space in the same orbit as before and would show up later as meteor showers when the bits and pieces of the nucleus encountered the earth's atmosphere. So the dirty snowball model was consistent with Schiaparelli's correlation between comet orbits and meteor streams.

What about the gas jets from the nucleus? The Whipple model accounted for those too. There must be spots on the surface of the nucleus where ice predominates over dust. Perhaps meteoroid impacts have knocked out craters and exposed fresh ice, or large chunks of the nucleus have split off under gravitational or thermal stresses and bared broad stretches of the comet's frozen insides. Those icy patches would give off localized jets of gas when warmed by sunlight.

Perhaps the most impressive feature of the Whipple model was its success in explaining the unpunctuality of comets such as Encke's, which had cast doubt on Newton's work centuries earlier. The dirty snowball model demonstrated that comets really did obey Newton's laws after all. What confused earlier astronomers was the fact that the law of universal gravitation was not the only Newtonian principle acting on comets. Also involved was his famous third law of motion: For every action there is an equal but opposite reaction. Here is how that law enters the picture. Comet nuclei spin around, toplike, on an axis of rotation, as the earth does; and just as a given point on earth (say, Chicago) faces sunward for part of the day and away from the sun for the remainder, the gas-releasing icy patches on a comet's face alternate between "day" and "night" as the nucleus whirls. Sunlight shining on those patches warms them and makes them release vapor while they face toward the sun. When facing away from the sun, they cool down, and gas emission falls off.

Gas release is an action. Therefore it has an equal but opposite reaction, namely to push the nucleus back in the opposite direction from the escaping gas. This process is identical to the one that makes a skyrocket fly. Someone lights the fuse on the rocket, and after a few seconds' delay the fuse burns down to the propellant and ignites it. A jet of heated gases rushes out the rear of the rocket, and the equal-but-opposite reaction described by Newton pushes the rocket forward. On a comet, the stream of gas comes from warmed ice rather than from rocket fuel, but the effect is the same. Whichever direction the gases take, the nucleus gets pushed the opposite way.

Now things start to get a little complicated, partly because of the nuclear spin and partly because the ices on the comet's surface do not flash into vapor immediately when the sun's rays touch them. It takes a while for the frozen gases to soak up enough heat to vaporize themselves. To illustrate, put an ice cube outside in the sunshine on a hot day. The cube does not melt at once into a puddle of water. The melting process takes several minutes.

So it is on the comet nucleus. If sunlight hits a patch of ice at "dawn," just as the sun is coming up over the comet's "horizon," then it may be "noon" on the comet before the molecules in the ice start to sublimate away. Like many of us, those molecules are a little slow to get started in the morning. (This delay before sublimation corresponds, in a sense, to the burning time of the fuse before a skyrocket takes off.)

Does this mean that comets always push themselves away from the sun, since the vapors escape only on the sunward side? Not necessarily. Remember, there's another factor involved here too: the spin of the nucleus. Which way the nucleus rotates will govern how the rocket effect of escaping gas alters the comet's course.

Not all nuclei spin in the same direction. Some rotate clockwise, others counterclockwise; and the direction of spin determines whether the reaction from the escaping gases will delay the comet's return or hasten it. Suppose the comet is spinning on its axis in the opposite direction to its path around the sun, so that the orbit is clockwise and the nucleus spins counterclockwise. This means gases will sublimate away with peak force late in the comet's "afternoon," when they are escaping approximately in the opposite direction to the comet's forward motion in orbit. As a result, the sublimating gases will "put the brakes on" the comet and slow it down slightly, much as astronauts in the space shuttle fire the engines to decelerate their craft prior to reentry. The braking effect on a comet is less dramatic; its gas release is gentle by comparison to a burst from the shuttle's engines. Too, the comet nucleus may be small by astronomical standards but is nonetheless a tremendous mass with great inertia (resistance to movement), so it gets pushed only slightly out of orbit by the sublimating gases. The nucleus moves perhaps a thousandth of one percent closer to the sun. That is roughly equivalent to half an inch out of a mile. Such a change in orbital radius is far too tiny to detect from earth. The only measurable difference shows in the comet's period. It loses a few minutes because its orbit is ever so slightly tighter now.

Slowing down a comet, then, will actually make it return more quickly. Is this a paradox? Not at all. As we saw earlier, it is the natural outcome of Kepler's laws of planetary motion. Decelerate a comet, and it drops into a smaller orbit with a shorter period. Accelerate a comet, and you send it into a higher orbit with a longer period, so that the comet takes more time to return.

This is why Encke's Comet kept showing up ahead of schedule. The rotation of its dirty snowball nucleus—in a direction opposite to that of its orbit—produced a rocketlike braking effect that nudged the comet into a briefer orbit on every return and kept shaving time off its period. (The braking effect seemed to grow weaker over the years because the axis of rotation was tilting steadily, and the change in inclination caused it to expose less of its icy surface directly to the rays of the sun. Therefore, less ice sublimated, the rocket effect diminished, and Encke's Comet settled into a more predictable schedule.) If a comet spins in the same direction as its orbit, the opposite effect occurs. The gas release accelerates the comet, pushing it into a higher orbit with a slightly greater orbital period, so that it is a little slower to return the next time. These and other cometary phenomena were hard or impossible to explain using the flying sandbank model, but the dirty snowball model suited the known facts about comets ideally.

Oort's Comet Cloud

About the time Whipple was developing the dirty snowball model, one of his colleagues in Europe, the Dutch astronomer Jan Oort of the University of Leiden, was working on the questions of where comets come from and how they originate. Like other astronomers before him, Oort was tempted to tie in the genesis of comets with the gap between the orbits of Mars and Jupiter. Had a planet actually existed in that space long ago and broken up somehow, sending a shower of cometlike bodies shooting out through the solar system like the shrapnel from an exploding land mine? It certainly appeared that way at first. The zone between the orbits of Mars and Jupiter is occupied by a swarm

of several thousand small rocky bodies known as the asteroids. The biggest of them, Ceres (the search for which inspired Gauss to invent that orbital motion equation that so
helped Encke), is about the size of Texas or Montana. Altogether the asteroids have something like one percent the
total mass of the earth, and they look like the wreckage left
over from the destruction of a planet—the same event, Oort
thought at first, that might have given birth to the comets.

Oort had to abandon that idea in view of the makeup of
asteroids. They are stony or metallic bodies similar to terrestrial rocks, unlike the icy nuclei of comets, which appear
to be made up mostly of frozen water and carbon dioxide.
Suggesting that asteroids and comets originated in the
same event, then, was like postulating a common origin for
bricks and snowballs, and Oort saw he would have to look
elsewhere for the "comet factory." If comets did not originate in the inner solar system, he reasoned, then they must
come from somewhere farther out. In that case, one would
count on finding a population of comets with highly elongated orbits that carried them in from thousands of astronomical units away; and that is just what Oort discovered.
He checked his data on comet orbits and found a population
of comets with orbits stretching out to twenty thousand astronomical units (almost two million million miles) or more.
Later studies uncovered other comets with orbits reaching
out more than seven times that far, into a great comet "reservoir."

Out in the dark depths of space beyond Pluto, it looks
as if a vast cloud of comets slowly circles the sun. This
comet swarm is now known as the Oort cloud. It is thought
to extend about halfway to the nearest star and occupy a truly mind-boggling volume of space: approximately four thousand billion billion billion billion,
or 4,000,000,000,000,000,000,000,000,000,000,000,000,000
cubic miles. A figure like that is hard for the human mind to
grasp, so let's put it into more easily understandable terms.
If you tried to build a scale model of the Oort cloud out of
table salt, and let each grain of salt stand for the volume of

the earth, then you would wind up with a ball of salt more than two miles wide.

The Oort cloud, like the comet nucleus, is as yet unseen, so we can only make educated guesses about its statistics. Most astronomers think the Oort cloud contains about 100 billion comets, but the cloud's total mass is astonishingly small—only a tenth to a hundredth that of the earth, or approximately the same as the mass of Mars or our moon. Seen from the outer edge of the Oort cloud, the sun would appear to be merely one more star in the galaxy.

If comets originate so far away from the sun, then something must kick them toward the inner solar system. The Estonian astronomer Ernst Öpik suggested that passing stars perturbed the orbits of comets in the Oort cloud and put them into new orbits that would take them in close to the sun. This mechanism fitted in with Oort's model so well that the far-flung swarm of comets is now sometimes known as the Oort-Öpik Cloud.

How the Oort cloud formed is still a matter of debate, but most astronomers think the comets condensed out of the same primordial gas and dust from which the sun and planets arose. One theory says that molecules condensed around dust particles, producing tiny ice-coated bits of dust that gradually came together to form the big dirty snowballs of Fred Whipple's model.

Another, slightly different, theory maintains that the comets condensed from gas and dust blown out of the inner solar system at a time, many millions of years ago, when the solar wind blew much more fiercely than it does now. Our sun is suspected of being a T Tauri star, or one that puts out more light and particle radiation at some times than at others, and many astronomers think the sun once went through an especially "windy" phase that blew much of the light material (gas and dust) from the inner system out to the realm beyond Jupiter, where temperatures were low enough for water ice and other chemicals to freeze around dust particles and get the comet-forming process started.

Or did comets really form from the same cloud of matter

as the sun and planets? One new school of thought about the origin of comets is based on quite a different assumption. Maybe the comets are not part of our system's primordial nebula at all. It is just possible that they were born from "alien" gas and dust at the edge of our star system. The sun and planets are orbiting the center of our galaxy, and from time to time our system's course takes us through one of the great spiral arms of dust and gas that radiate out from the galaxy's hub. Those dust and gas clouds are not very dense (indeed, they constitute a better vacuum than any we can generate in laboratories) but have such tremendous volume that they also possess a tremendous mass, and that means gravitation—enough, some astronomers think, to allow those great nonluminous cloudbanks to sweep away much of the Oort cloud every time our system passes through one of them. What the galaxy taketh away, however, it also giveth back. The dust and gas in those clouds may itself condense to form a new population of comets, replacing those it wiped away.

This is not a peaceful process. Many of the comets in the initial Oort cloud get scattered (according to this model) and diverted into orbits that will carry them sunward to become long-period comets like Kohoutek. Later their periods may be whittled down if they stray too close to Jupiter or Saturn, and in time some of the comets may settle into tight orbits around the sun, in the manner of Encke's Comet.

The Krypton Model

Then again, one radical school of thought says that Oort was right the first time when he sought to explain the comets' birth in terms of a planet's destruction. One exponent of this view is the Canadian astronomer M. W. Ovenden, who has postulated that perhaps there really was a planet once between the orbits of Mars and Jupiter. This planet, dubbed Krypton in honor of Superman's homeworld, is Ovenden's candidate for the maker of both the comets and the asteroids. Ovenden imagined that Krypton had about 100 times the mass of the earth and was destroyed approximately ten million years ago—almost yesterday on the scale of geological time—by some as yet unspecified disaster. The death of Krypton, Ovenden suggested, cast out the comets like shrapnel from an exploding grenade.

This scenario is unlikely. Had anything of the kind happened ten million years ago in the inner solar system, the earth and moon and Mars would have been pelted with bits of Krypton, and the rate of cratering on those worlds would have risen sharply. There is no such evidence in the record of the rocks. For the last sixty million years or so, large meteorites appear to have been hitting us at a reasonably constant rate, with no dramatic peak around ten million B.C.; and the estimated rate of cratering for the moon and Mars roughly equals that of the earth.

The Krypton model still has its defenders but has been modified to better suit the geological record. Instead of blowing up in a sudden, catastrophic event, it seems more likely that Krypton (again, assuming it really existed and was destroyed) broke up more gradually, perhaps under gravitational stress from nearby Jupiter, which then would have absorbed more than 99 percent of its late neighbor's mass and left only a scattering of debris—what we call the asteroid belt—to mark where Krypton had been.

Why argue for Krypton's existence at all? There is very little evidence for it, but that evidence is intriguing, and the Krypton model is consistent with what we know about the distribution of long-period comets. Several dozen of the more than one thousand comets identified so far have orbital periods of about ten million years. These all appear to be first-time comets that had never entered the inner solar system before. Are they flying chunks of Krypton that only recently returned to the place of their birth?

If these comets were cast out of Krypton, their orbits would tend to cluster close to the ecliptic (the plane of the hypothetical Krypton's orbit) and the point on the ecliptic, known to astronomers as its ecliptic longitude, where Krypton stood when it was destroyed. There should also be another such concentration directly opposite that point on the celestial sphere, the imaginary "globe" of the heavens as seen from earth. Sure enough, we find a concentration of long-period comet orbits around two opposite points near the ecliptic, one in the vicinity of 260 degrees ecliptic longitude and another at a spot directly across from it on the celestial sphere, near 80 degrees ecliptic longitude. This pattern is understandable if these long-period comets—supposedly the fastest-moving fragments of the planet to go into orbit around the sun—were ejected in Krypton's death throes and are just now returning to their point of origin. To use an analogy: If you stood on the earth and used an extremely powerful blunderbuss to shoot a cloud of pellets into orbit around the sun, the pellets would come back around to their point of departure in much the same pattern as that of the long-period comets. The gravitational fields of other celestial objects would have perturbed their orbits slightly and introduced some "scatter" into their initial tight concentration, but the pellets would still show up again near their point of departure.

So perhaps there is something to the Krypton model after all. Right now it is definitely a minority view among astronomers, but so was the Copernican system in the not-too-distant past. Maybe someday we will find proof that these visitors from deepest space actually launched themselves from our own backyard. For the moment, however, most evidence rests on the side of the Oort model. Let us pick a comet from Oort's swarm and ride it, in our imaginations, as the comet swings sunward.

SIX

A Comet Ride

CHOOSING A COMET

The best choice for a comet ride is a medium-sized one that is still active enough to demonstrate its "vital functions" at work and that has an orbit similar to Halley's Comet. Such an orbit will take you on a tour of the solar system, culminating in a swift dash past the sun.

Now that you have your comet, place yourself in an imaginary spacecraft on an interception course with it near its aphelion in the outer reaches of the solar system. Even with radar to help, it is hard to pick out your target, for a mile-wide snowball, though huge by everyday human standards, is virtually nothing in the immensity of space. If you were not looking intently for the comet, and if you did not know almost exactly where it would be, you would stand very little chance of ever finding it. Yet there it is: a blip on the radar, just ahead. A gentle thrust from your vehicle's engines and you start moving up on your frigid quarry.

Radar is an absolute necessity for navigation in this case, partly because the distances involved are so great, and partly because visible light is hard to come by. Here in the dark "suburbs" of our star system, the sun is no longer the blazing disk familiar to earth dwellers, but only a point of light amid the other stars. It looks about as bright from this distance as Venus does to observers on earth. Since the sun sheds so little light, you will have to use searchlights to illuminate the comet's surface.

Your first view of the comet is not attractive. At first it looks like exactly what it is: a great dirty snowball. Roughly spherical, it has a muddy-looking gray-brown surface dotted with irregular splotches of pale ice. One side is considerably darker than the other. It looks rather mangy, and it is hard to believe that this dull object will soon create one of nature's most beautiful displays in the nighttime skies of earth.

The comet has a measurable gravitational field, but not much of one. If you dropped this book from shoulder level above its surface, the book would take about half an hour to fall. The surface is made up of water and carbon dioxide ice, plus silicate dust and rock in diameters ranging from a few millionths of an inch up to baseball size. Other components are merely traces. The whole mass is much like the permafrost, or permanently frozen ground, of northern Canada and the Arctic USSR, only more porous and crumbly.

Seen close up, the comet's face resembles a gigantic cinder or a lump of frothy volcanic rock enlarged a million times. Meteoroid impacts have caused some of the irregularities; the comet is pockmarked by thousands of craters. Some are only a fraction of an inch wide, while others measure up to a hundred yards. Amid the craters are the holes blown out by the escape of gases from inside. As the nucleus nears the sun and starts to heat up, frozen gases within the nucleus sublimate and rush out through openings in the surface, at velocities of hundreds of yards per second. The steady stream of escaping gas scours out pores in the "skin" of the comet. But it will be a while yet before the comet reaches that point in the inbound leg of its orbit.

For the moment things are quiet on the nucleus, so there is plenty of opportunity to do some "prospecting." Leave your ship and maneuver yourself over to the comet's surface. Be careful not to accelerate too much, for your spacesuited body, though weightless, still has every bit as much mass as it would have in the gravitational field of the earth. Therefore you will build up just as much momentum on the trip over to the comet as you would in an equivalent journey on earth, and your impact will be just as painful if you land faster than a few inches per second. So take your time. The comet won't go anywhere without you—for the time being, at least. Fortunately this comet has virtually no spin on its nucleus, so landing here will be much easier than on a rapidly spinning body.

Once you set foot on the comet, you had best take precautions to keep yourself there. In such a weak gravitational field, a slight push with your hand or foot would suffice to send you flying off into space. So drive a piton or two into the comet's surface and tie a line to the piton as a mountain climber would. You are actually climbing a mountain as big as many in the Rockies and the Cascades. This mountain just happens to be flying along in orbit, and the danger here is not the insistent pull of gravity but the relative absence of it.

Having anchored yourself securely, take out a hammer and chip away gently at the surface of the nucleus. The minerals fly away in tiny bits, almost like a cloud of dust, except that the dust particles do not form a cloud as they would on earth. There the atmosphere keeps dust particles in suspension. Here there is no air to do so, and the little bits of rock fly slowly out into space, each on its own separate trajectory.

The comet's "soil" is light and porous. You can sense its texture through the hammer's handle and the vibrations it transmits through your bones. It feels as if you are hacking away at densely packed snow or plastic foam. Keep digging until you reach the ice. That pale mass of frozen gases is a legacy from the distant past. You are looking at a bit of cosmic history: ice crystals that formed before the human race arose and perhaps before the first snow ever fell on earth.

This is a special kind of ice. Chemists call it a clathrate hydrate. A hydrate is a water-bearing compound, and clathrate means "latticed." In this case the water molecules have linked up in a three-dimensional lattice, and lodged in the spaces of this lattice are other kinds of chemicals, from nitrogen and carbon to the more complex hydrocarbons and other relatively big molecules, which are released into the tail and can be detected by spectroscope from earth. Visualize a clathrate hydrate in terms of a wine cellar. The nonhydrous molecules here stand for the bottles of wine, and the water molecules form the rack in which they are stored. Of course, the structure of a clathrate hydrate is actually more complex than this simple analogy indicates: The crystals take the form of dodecahedrons, or twelve-sided polyhedra, with the water molecules arranged on their corners and edges and the "foreign" atoms nestled in their hearts. The wine rack analogy is often used in college astronomy courses, however, to illustrate the storage of molecules in a comet's hydrate ices.

Driving in pitons as you go, make a circuit of the nucleus. Here and there you can see damage done on previous circuits of the sun. Great fissures run across the surface where gravitational stress and the pressure of expanding gases have begun to pry the nucleus apart. One of those fissures has already given way. A broad, bright surface of freshly exposed ice shows where a chunk of the comet roughly a hundred yards wide has split off during an earlier solar encounter. What exactly happened to that splinter of the comet is hard to say, but most likely it evaporated under the terrific heat of the sun near perihelion and left only a shower of dust and rock to indicate where the chunk of comet had been.

Will the comet hold together on this circuit of the sun? Time alone will tell, and it need not be a long time, for in the imagination you are free to speed up the comet's motion and thus compress the thirty-odd years until perihelion into a few minutes. This breakneck ride through the solar system will leave little time for sightseeing, as the planets slip past like highway signs on a turnpike. So watch carefully as the comet whizzes from the outer to the inner solar system and back again; and note how dramatically comets interact with the rest of the system.

Comets and the Planets

The outermost planets—Pluto, Neptune, and Uranus— are shadowy unknowns at the edge of our astronomical knowledge. What a contrast with Saturn, the most spectac- ular world in our solar system! Saturn's gravitation can throw comets off their expected schedules by hours or even days. The effect of Saturn's gravitational field on comets' or- bits was once the bane of astronomers trying to calculate when periodic comets would return. (Remember the ruined constitution of poor M. Lepaute.) The great "magnet" for comets is not Saturn, however, but its colossal neighbor Ju- piter (Fig. 6-1), whose gravitational field was responsible for capturing this particular comet thousands of years ago and swinging it into a periodic orbit about the sun.

There was a time when astronomers thought Jupiter it- self was a comet source. Recall how Hevelius imagined com- ets being "burped" out of the Jovian moons and then spinning through the void like cosmic tiddlywinks. In a cer- tain sense this view of how comets originate was correct, for Jupiter does send many comets our way, though not by cast- ing them out directly. Jupiter's steep-sided gravity well acts as a "funnel" for comets and can transform even an ultra- long-period comet—one that returns only once every million years or more—into a frequent and regular visitor to the in- ner solar system.

The capture process works as follows (Fig. 6-2). A long- period comet moves in from the suburbs of the solar system. For purposes of illustration, let us assume the comet passes Jupiter on the left at a distance of a few million miles. Both Jupiter and the sun are on the right, from the viewpoint of an observer on the comet. This means the comet, on its ap- proach, circles the sun in a clockwise direction as seen from a vantage point above the sun's north pole.

Even at such a distance, Jupiter's mighty gravitational

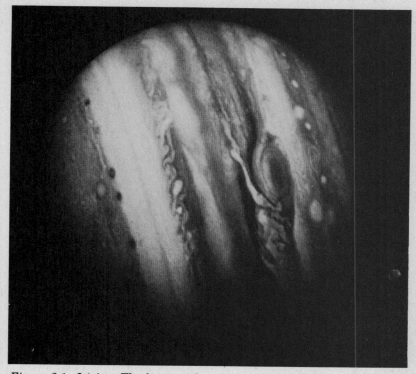

Figure 6-1. Jupiter. The largest planet in our solar system, Jupiter captures many comets and has created a whole comet "family" for itself.

field sways the comparatively tiny comet off its original course and whips it about to starboard. Jupiter's gravitation is not sufficient to place the comet in orbit about Jupiter but does send the comet into a new elliptical orbit, with a much shorter period, around the sun. This former long-period comet has become a denizen of the inner solar system now as part of Jupiter's comet family. The comet also has had the direction of its orbit reversed. Previously it followed a clockwise orbit around the sun. Now it orbits counterclockwise, approaching the sun from the right-hand side and departing on the left.

Stretch your imagination, and you might be able to see Jupiter's gravity well, an indentation in the fabric of space

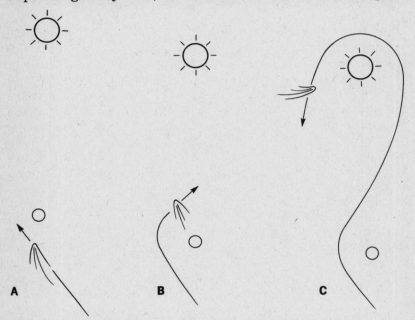

Figure 6-2 A, B, and C. How Jupiter Captures Comets. Here we see a comet approaching the sun in a clockwise orbit. As the comet passes close to Jupiter, the giant planet's gravitational field swings the comet around into a new, counterclockwise orbit around the sun.

with the planet's huge banded mass at the well's very bottom. Directly ahead, the sun's gravity well stretches out ahead of the comet as the nucleus slides down the sides of the well, on its way to a brief solar rendezvous.

Through the Asteroid Belt

Jupiter falls behind. Now look carefully, and you may be able to see a few pinpoints of light between the orbits of Mars and Jupiter. These are the asteroids, the swarm of planetesimals, or tiny planetlike objects, that played such an important part in the development of celestial mechanics and consequently in the history of comet studies as well. Your passage through the asteroid belt holds very little danger. Although several thousand known asteroids share the space between Jupiter and Mars—and there may be dozens of smaller, undetected asteroids for every one known to astronomers—your chance of colliding with one of them is extremely slight. The asteroids themselves are so tiny, and the spaces between them are so enormous, that your dash through the belt is likely to be collision-free. All the same, collisions can happen among the asteroids; and one is about to occur right now.

Unknown to you, a chunk of rock roughly the diameter of a meatball has been speeding through space on an interception course with your comet. The odds against a collision are more than a billion to one, yet the little rock finds its mark. It hits the nucleus just over the horizon from your position, blasting out a crater roughly a yard in diameter and sending chunks of rock and ice spewing out in all directions. Looking up and ahead, you can see a fountain of particulates shoot up from the surface. You are lucky that the incoming rock landed no closer to you. If flying debris from the impact had holed your suit, you might have died in a moment from explosive decompression as the air inside escaped to space. In the imagination, however, you lead a charmed life.

Now to inspect that crater. It is about a foot deep and stands out in sharp contrast to the surrounding surface, for the impact has knocked away dusty material and exposed fresh, gleaming ice underneath. Some bits of ejecta (tossed-out material) from the impact site fly off into space at velocities that will take them out of the comet's shallow gravity and send them spinning into their own orbits around the sun. Someday they may be drawn into the gravity well of the earth and immolate themselves as meteors in the evening sky. Other, smaller chunks of ejecta remain trapped in the comet's gravity well and will become the comet's own satellites—but only for a few hours, because those tiny bits of comet debris are about to be swept away into deep space by a dramatic change in the comet.

The Coma and Tail Appear

While you are preoccupied with the crater, the comet is undergoing a dramatic transformation. As you approach the sun, more and more solar radiation falls on each square inch of your comet's surface. That radiation is absorbed, turns to heat, and warms the nucleus. Its surface is still far colder than the iciest winter temperatures in London or New York (only some spots in Antarctica ever approach the chill of a comet's nucleus), but even the faint added warmth from sunlight suffices to turn ice into vapor, and a cloud of very rarefied gas starts to spread outward from the nucleus.

The cloud is practically invisible from where you stand, but on earth the vapor reflects the sun's rays and simultaneously glows—faintly at first, then more and more brightly—with a luminosity of its own. The rays of the sun are striking gas molecules in the comet's "atmosphere" and making them fluoresce like the molecules in signs on Broadway. Those signs, however, advertise only mundane things like motion pictures and meals. The light from your comet's molecules is advertising a visitor from the depths of space, and the world takes notice. Everywhere on earth that there are telescopes, astronomers look skyward and focus on the rocky iceball you are riding. You and the nucleus of course are invisible from the distance of earth, but the coma of your comet shines more and more prominently all the time.

Now the pressure of the solar wind starts to blow gas and dust backward, away from the sun, and your comet sprouts a recognizable tail. This one will be a beauty. The nucleus is giving off several distinct gas jets—including one from the crater you saw blasted out earlier—and promises to show a spectacular set of tails.

Look around, and you note fountains of gas shooting up like rarefied geysers. With them arises a cloud of dust particles that streams away with the tail gases. Solar radiation is doing more than making the molecules around you shine. It is also breaking them apart into smaller pieces. Like clay pigeons, the water and carbon dioxide molecules in the coma and tail are getting shattered into hydrogen, hydroxyl, carbon monoxide, atomic carbon, and other submolecular debris, and that wreckage is drifting away under the impetus of the wind from the sun.

Some of the dust liberated from the nucleus is also drifting away into space—but not all of it. Some of the larger particles are too heavy to be blown away by the solar wind, and they remain trapped in the gravitational field of the nucleus. The particles sail slowly upward from the surface, travel on curved suborbital paths about halfway around the nucleus, and come to rest on the opposite side from which they started. Since this comet has very little spin on its nucleus, the result is a nucleus with one side much darker than the other, the dark side having been produced by the slow accumulation of particles over the centuries.

Comets and the Martian Atmosphere

By now the asteroids are well behind. You are crossing the orbit of Mars. Take a good look at the red planet as you pass, for its face may bear witness to comet impacts millions of years old. Look specifically at the Martian polar caps. They stand out, brilliantly white, against the reddish dust and rocks. The polar snows contain much of what used to be the air and seas of a living, active Mars: not the dead world we see there today, but a dynamic planet where free-flowing water carved stream- and riverbeds like those on earth. Their sinuous channels showed up plainly on Mariner photos of the red planet and told scientists on earth of a time when Mars had its own Seines and Mississippis. The water that created those landforms is now locked up as ice in the polar caps. Liquid water cannot exist on the Martian surface today, for temperatures and the surface atmospheric pressure are too low. The Martian atmosphere is so thin that taking a walk on the Martian surface would be little different from walking in space (except for the gravitation of the planet underfoot), and an open container of liquid water would evaporate in a matter of seconds, then fall out of the air as fine ice crystals.

We know that the ancient atmosphere of Mars contained large amounts of water and carbon dioxide, for those are the principal components of the polar caps—and of comets as well. Did comets therefore give an atmosphere to Mars in the distant past? The traditionally accepted model of the evolution of Mars's atmosphere is the "outgassing" model, in which the Martian air is assumed to have formed from gases seeping out of the planet's interior shortly after Mars was formed. Yet comets offer another tenable model that neatly explains both the origin of the Martian atmosphere and the chemical composition of the polar caps.

Comets colliding with Mars would have vaporized themselves on impact and left craters. Large portions of the face of Mars are heavily cratered, and even a mere handful of comets could have provided enough carbon dioxide, water vapor, and other gases to form an atmosphere thick enough for clouds and rain. Some astronomers, then, suspect that the ancient atmosphere of Mars was not the result of "outgassing" but instead was the product of falling comets that gasified themselves on impact and thus gave the planet an atmosphere, back when Mars was a much warmer place than it is now. But what chilled the planet's climate and froze the Martian atmosphere and seas is still unclear. Possibly unmanned probes or astronauts from earth will one day determine what turned our neighboring red world into a dead one.

Did Comets Make Life on Earth Possible?

Ahead of your comet now is the earth, white and blue against the black backdrop of space. Mars and the earth are so similar in many respects that one can hardly help wondering: If comets provided Mars with an atmosphere, might they have done our world the same favor? That hypothesis is consistent with much of what we know about the history of our world, including the chemistry of its primeval air. The earth has not always had its present oxygen-rich atmosphere. The oxygen we breathe was once only a trace element in the air, compared to much greater amounts of carbon dioxide and water. Large quantities of free oxygen came much later, after green plants evolved with the ability to convert carbon dioxide into carbohydrates (that is, sugars and starches) and oxygen. The plants had no need for oxygen. To them it was a toxic waste product, so they cast it away into the atmosphere, where the oxygen accumulated and eventually made possible the evolution of oxygen-breathing animals like ourselves.

Those early plants evolved in an environment with abundant carbon dioxide and water, chemical clues that comets may have contributed heavily to the atmosphere and the hydrosphere of the ancient earth. So the next time you drink a soda, reflect that part of the water and carbon dioxide in that carbonated beverage may have been delivered here by comet impacts before life on earth began.

The Solar Wind

At this moment an invisible war is going on between the gas output of your comet and the particle output of the sun—the solar wind. The leading edge of the comet's "atmosphere" collides with the outward-moving "atmosphere" of the sun, and the result is a blunt, rounded "shock front" somewhat like the bow wave of a ship, preceding the comet through space. As the comet passes, it leaves behind it a turbulent wake of wobbly particle streams to mark its passage through the solar wind.

Most of the time this silent battle between the comet and the sun goes on unseen to human eyes. Right now your comet, however, is about to make this subatomic combat visible in a spectacular way. The solar wind is "strung" with magnetic force lines that run at angles to the comet's orbit. In some places the magnetic field reverses itself, and the result is what physicists call a sector boundary, an invisible line on either side of which are opposite magnetic fields. The sector boundary is comparable to the highway divider that separates northbound from southbound traffic. Your comet is coming up on a sector boundary just ahead. When the comet hits the boundary, strange things can occur—as you are about to see.

The sector boundary has very little material existence. It does nothing to slow down the nucleus. The reversal in the magnetic field has dramatic effects on the tail, however, and these may be best understood by invoking an image from sports. Imagine a tennis player with superhuman strength attaching a set of crepe-paper streamers to a tennis ball and then lobbing the ball toward the net with all his might. The ball hits the net, pushes it out of shape, and finally rips through, leaving its streamers stuck in the net.

Much the same thing happens when your comet encounters a sector boundary. The ball stands for the comet, the net for the sector boundary, and the paper streamers for the comet's tail. When the comet passes through the magnetic force lines at the sector boundary, those lines bend out of shape as the net does in that imaginary tennis serve. The comet rips through the boundary, but as it passes, the force lines come together again in the comet's wake and pinch off the tail, which is left hanging in space while the nucleus sails on ahead of it, suddenly tailless.

This tail-cutting phenomenon is known as a disconnection event, and it is going to cost your comet its tail temporarily. Observers from earth watch with interest as the invisible force lines snip off the tail and set it adrift in the void. The loss of a tail, however, is no great setback to the comet. It simply starts growing another.

The gas and dust in the detached tail will meet two different fates. The gas will be blown away quickly by the solar wind, but the dust will hang around longer and show up in terrestrial skies for a long time afterward as the zodiacal light, a glow often seen on the horizon just before sunrise and following sunset. The zodiacal light is most prominent in the tropics and consists of sunlight reflected off the dust particles left behind by comets.

Perihelion

The comet is moving more and more rapidly now, in obedience to Kepler's laws. Venus and Mercury are coming up fast. The thick clouds of Venus make it impossible to see the planet's surface, but tiny, sun-baked Mercury is naked to view. On this airless little world you can see, on close examination, the signs of comet impacts (Fig. 6-3). Some of the craters on Mercury are surrounded by light-and-dark swirl patterns similar to those seen in pound cakes. These patterns are the "calling cards" of comets that slammed into the Mercurian surface. Comet dust freed on impact settled to the ground in intricate patterns as the nuclei exploded into vapor. There the dust remains and probably will remain for as long as Mercury endures.

Perihelion! At this point the heat and hard radiation from the sun are so intense that no one could survive except in an imaginary journey. Observers on earth can no longer see the comet, for it is hidden behind the solar disk. Therefore you will be the only witness to the comet's breakup and destruction.

This near the sun, the thermal and gravitational stresses are too much for the comet to bear. It starts to disintegrate. First it shudders, then abruptly splits in half. It breaks up silently, for here there is no air to carry sound. Even in silence, however, the sight is fantastic.

The two halves drift slowly apart, then they too divide,

Figure 6-3. Mercury. The innermost planet of our solar system has been bombarded frequently by comet nuclei, as its heavily cratered face reveals.

like cells reproducing. At first, thin and glowing bridges of cometary material connect the pieces. Then the bridges dissipate, and the four chunks of ancient ice and dust begin their long climb up and out of the solar gravity well. When the comet reappears on the sun's far side, watchers on earth will see a multinucleate comet, its fragments hurtling through space on similar but now separate orbits.

The breakup was not a neat one. Like a dry cookie, the comet shed many smaller fragments as it crumbled. Those crumbs of the nucleus may vanish forever into outer space, perish in their next encounter with the sun, or destroy themselves in collisions with other celestial bodies. We know in detail what happens in such collisions because earlier in this century a small chunk of a comet struck our earth—with catastrophic results.

SEVEN

The Siberian Comet Mystery

"A TONGUE OF FIRE"

On the morning of June 30, 1908, something fell from space into the atmosphere over Siberia. The object raced northward across the sky, trailing a cloud of vapor behind it and making a noise so intense that observers on the ground below clapped their hands to their ears in pain. What happened next is best described in the words of an eyewitness:

> I was sitting on my porch facing north when suddenly, to the northwest, there appeared a great flash of light. . . . [the] shirt was almost burned off my back. I saw a huge fireball that covered an enormous part of the sky [and then] I felt an explosion that threw me . . . from the porch. I lost consciousness.

Other eyewitnesses reported that the falling object was "pipe-shaped," meaning cylindrical, and shone so brightly that it was impossible to view directly. A towering cloud of black smoke was said to have risen from the spot where the fiery thing landed. At least one report mentioned a "tongue of fire" that emanated from the cloud. The explosion was accompanied by a noise like that of powerful artillery, and the ground shook as in an earthquake, scaring the engineer on a Trans-Siberian Railway train so badly that he halted the locomotive lest it be derailed. An atmospheric blast wave swept across the landscape, blowing carpenters off the scaffolding on a house and collapsing tents. Livestock panicked and scattered. Later, herdsmen found peculiar scabs on the hides of their animals, as if some unfamiliar kind of energy had burned them.

Reports of this disaster took some time to filter out of Siberia, for the incident took place in an isolated and underpopulated corner of Russia, and communications then were much slower and less reliable than they are now. By the time the news reached St. Petersburg (later renamed Petrograd and now known as Leningrad) and other major cities in western Russia, the story had been grossly embellished. Details had been added and exaggerated until very little accurate information about the incident remained. According to one report printed on a calendar in St. Petersburg, the falling object landed close to a railway line. Passengers on a train had had a chance to inspect the fallen body, a light-colored chunk of stone roughly two thousand cubic feet in volume, after it cooled down from its initial red-hot state.

The tale on the calendar was pure fiction. The object had fallen nowhere near a railroad, nor had anyone reported finding any pieces of the thing, whatever it might have been. This calendar yarn turned out to be an important document in the history of science, however, for in 1921 it came to the attention of a scientist named Leonid Kulik at the Mineralogical Museum in Petrograd. Kulik's investigation of the 1908 Siberian event would bring it to the attention of the entire world and set off a scientific detective story that eventually would lead researchers back to the discoveries of eighteenth-century comet hunters.

Leonid Kulik's Quest

Leonid Kulik was anything but an ivory-tower scientist. A former revolutionary, he had been arrested for subversive activities under the tsarist regime and had served as the Russian equivalent of a forest ranger before he developed an interest in geology and became one of the Soviet Union's leading mineralogists. Meteorites were his hobby. Kulik had helped assemble a large collection of them at the Mineralogical Museum. He saw the calendar report as he was preparing to go on a meteorite-hunting expedition, and he figured this might be an opportunity to add a truly stupendous specimen to the museum's collection.

First he checked old newspaper reports of the Siberian disaster. They were less than enlightening, but the press accounts agreed that something very luminous had fallen out of the sky over Siberia in the spring of 1908 and done great damage around its point of impact. No specific impact site was mentioned, so Kulik had to try to work it out for himself. His research indicated that the impact site lay somewhere in the basin of Siberia's Stony Tunguska River, in a region of swamps and low wooded hills. In 1921 Kulik led an expedition to Siberia under the auspices of the Soviet Academy of Sciences and questioned locals about the disaster of 1908. Their answers were short on scientific detail but convinced Kulik that some great fiery object—most likely a large meteorite—had indeed fallen in the Siberian wilderness thirteen years before.

Unfortunately, Kulik's time was limited on this trip, and so he had no opportunity to scout further for the site of the object's landing. He and his colleagues returned to Petrograd, where Kulik spent the following several years gathering all the information he could about the 1908 disaster before trying again to find the alleged meteorite's grave. In 1924 some interesting stories filtered back to him through the research of a geologist named Obruchev who visited the Stony Tunguska River that year. Obruchev heard tales from natives about a holy place surrounded by a leveled forest, in the barren lands several days' travel north of the river. At that spot a fierce wind was said to have blown down the trees. The wind, Obruchev was told, carried such force that it knocked down buildings many miles away.

Had the natives any idea what caused the mighty wind? They blamed an angry god. That was definitely not a scientific explanation, so Kulik set out for Siberia again in 1927 to search for the meteorite. This time he had better information to guide him. Between his journeys he had been able to pin down the impact site more precisely, and now he knew to look at a spot approximately five hundred miles north of the town of Irkutsk, in the vicinity of Lake Baikal. The phenomena described by the eyewitnesses—the flash of light, the huge cloud, and so forth—centered on that location. Kulik had a superabundance of eyewitness observations to help him determine the impact site, because the object's fall was either seen or felt over an area roughly the size of Texas. By noting the position of the cloud and fireball relative to the various points of observation (northeast from Krasnoyarsk, northwest from Kirensk, etc.) Kulik narrowed down the possible area of impact to a few square miles of wilderness.

Getting there was easier said than done. Kulik and his team persevered, however, and finally reached the site. What they saw there had no precedent in science.

The Devastated Area

Over an area of several hundred square miles, full-grown trees were blown down like matchsticks in a radial pattern centered on the spot Kulik had determined for the object's fall. The trees were scorched as if by fire, and stripped of leaves and branches. Whatever destroyed them had ripped them up by the roots, knocked them over, and bombarded them with a sudden, fierce pulse of heat, all at the same moment.

Kulik assumed a meteorite could have done all this. The problem was, no signs of a meteorite impact could be found. A meteorite would have left a crater where it landed; yet there was no crater. The nearest thing Kulik found to a crater was an odd-looking set of concentric ripples in the ground, indicating that something had sent intense shock waves out through the soil from the center of the explosion. At the center of this set of rings, on the site that a later generation would call ground zero of the explosion, trees were still standing upright but had been driven partway into the ground, as if by a piledriver.

What did this strange assortment of damage? Kulik concluded that some highly unusual kind of meteorite had exploded just above the ground. The blast wave had hammered trees directly under the point of explosion straight down into the soil and had knocked down the surrounding forest in a pattern focusing on the heart of the fireball. The explosion was powerful enough to create those ripples in the soil and to knock down buildings more than forty miles away.

The absence of a crater, however, was a serious flaw in the meteorite-impact explanation, as was the absence of meteorite fragments. One would have expected to find bits of the postulated meteorite scattered around the impact site, but Kulik's thorough investigation turned up not a single chunk of meteoritic material. Nonetheless, Kulik had no better explanation for the Tunguska event, as it came to be called, and so he accepted the meteorite hypothesis. Kulik even dreamed up a mechanism to account for the midair blast. The explosion, he suggested, was caused by a falling meteorite or swarm of meteorites pushing a huge "bubble" of intensely hot gases ahead of them through the atmosphere. When the bubble burst, so to speak, it generated the fireball and blast wave.

Kulik wrote up his report on the 1927 expedition and delivered it to the Academy in December of that year. His story caused a sensation both in scientific circles and in the popular press, and Kulik gave a public lecture in which he described in frightening terms the possible effects of an impact such as the one in 1908: "had this meteorite fallen on New York, Philadelphia might have [had] only its windows shattered . . . but all life in the central area of the . . . impact would have been blotted out instantaneously."

Though Kulik clung to the meteorite explanation, the absence of meteorite fragments at the impact site continued to trouble him. He speculated, not very convincingly, that a large meteorite might have bounced off the earth's surface and back out into space, leaving no bits of meteoritic material behind. Kulik also imagined that the meteorite might have simply vaporized itself on impact, in which case no pieces much bigger than a molecule would have survived. The missing fragments puzzled astronomers as well, and they started looking for other possible explanations besides a meteorite impact.

The Comet Hypothesis

Several years after Kulik's 1927 expedition, a Russian astronomer named Astapovich proposed that the Tunguska event occurred when a comet fell into the atmosphere and exploded into vapor. Astapovich carried out calculations of the energy that an exploding comet might have released, and he decided that a comet colliding with the earth over Siberia could have unleashed more destructive force than a typhoon or a volcanic eruption. The comet hypothesis explained all the features of the 1908 calamity neatly, from the fireball and blast wave to the absence of a crater and meteorite fragments at the point of impact. An elegant hypothesis can never substitute for solid fact, however, and so debate continued about the nature of the Tunguska event right up to the date Russia entered World War II.

The war put a halt to Kulik's research. He was drafted and sent to help man the defenses of Moscow, was wounded in battle with the Germans in 1941, and taken prisoner. Several months later he died of typhus in a prison camp. Meanwhile an army officer named Alexander Kazantsev was running a military research and development center near Moscow and getting such good results that the Soviet government decorated him handsomely for his services. After the war, Kazantsev began a second and equally successful career as a science-fiction writer, and in one story he tried to explain the Tunguska event as the explosion of a malfunctioning nuclear-powered spacecraft from Mars. His suggestion received wide publicity but was not taken seriously by scientists, although some serious scientific explanations of the blast were scarcely more plausible. It was suggested that a tiny bit of antimatter fell from space and blew up with terrific violence when it encountered the "normal" matter in our atmosphere. When black holes came into fashion in astronomy, one of them was also invoked to explain the Tunguska event. Supposedly a minuscule black hole, no bigger than a pinhead but billions of tons in weight, had plunged through the earth and devastated Siberia with the resulting shock wave. As more and more evidence accumulated from studies of the Tunguska event, however, it began to seem that Astapovich's exploding comet provided the best possible explanation.

Some highly convincing evidence came from countless tiny droplets of fused rock and bits of magnetite, or magnetic iron oxide, that were found scattered through the soil at the site of the Tunguska event. These minuscule pieces of matter closely matched the chemical composition of "carbonaceous chondrites," bits of interplanetary debris that fall to earth occasionally as meteorites. A carbonaceous chondrite has about the same texture as a piece of very stale bread. It is highly friable, meaning you could crumble it between your fingers with no difficulty. You would get a pile of black or gray dust much like the samples found in the Siberian soil at the site of the Tunguska event; and so astronomers assumed that whatever blew up over Siberia in 1908 fell from outer space and had the chemical makeup and consistency of a carbonaceous chondrite, as well as a big supply of volatiles that would have expanded and caused an explosion when superheated by the object's descent through the atmosphere. Only one known celestial object has all these characteristics: a comet nucleus.

Suppose a comet nucleus had approached the earth unobserved. That was certainly possible. Vasili Fesenkov of the Soviet Academy of Sciences worked out the orbit of the Tunguska object and reported in 1966 that it must have moved up on the earth from behind the sun, so that the object was hidden in the sun's glare during its approach. (Fighter pilots during World War II used a similar technique to surprise enemy planes: They would attack from out of the sun so that the solar glare would hide them from view until it was too late for the target aircraft to escape.) Comets had followed such paths before. In 1957 a comet "sneaked up on" the earth in this fashion and was noticed only after it had rounded the sun and passed the earth's orbit on its way back into deep space.

Eight years later another event, this time over North America, put still more weight behind Astapovich's comet hypothesis. A meteorite fell into the atmosphere over Canada and exploded with a bright flash of light. No pieces of the meteorite were found—only a fall of dark dust much like that discovered in the soil at the site of the Tunguska event. This odd event, a scaled-down replay of the Siberian explosion of 1908, was almost surely the spectacular death of a comet fragment.

Today most astronomers concede that the Tunguska event was the work of a chunk of a comet. Most likely the fragment escaped detection because it swooped in from out of the sun and also because the fragment had lost much of its volatile part, so that it no longer gave off a coma and tail. The minicomet had just enough volatile matter left inside to create a terrific explosion when heated by air friction.

The Astapovich hypothesis was confirmed by a study of the mineral content of dust in a Siberian peat bog in 1977. The bog was located near the Tunguska site and was so isolated that it received mineral input only from the air. A layer near the surface of the bog was found to be full of tiny silicate spherules—that is, bits of natural glass—rich in heavy elements that are rare on earth but common in extraterrestrial objects like meteorites.

It took no great imagination to see what had happened. The comet fragment that blew up over Siberia in 1908 had vaporized itself, and the vapors had recondensed in the form of little glassy beads loaded with rare elements from the original nucleus. Then the silicate fallout settled to earth and was preserved in the Siberian soil and peat. One might say the microscopic glass balls wrote the comet's epitaph. They had been lying there for decades, waiting for scientists who could read their message.

APOLLO OBJECTS ARE DISCOVERED

While Astapovich was pondering the cause of the Tunguska explosion, astronomers were also talking about a new class of nonluminous but still cometlike objects that had been discovered by accident at a German observatory in 1932. An astronomer named Karl Reinmuth at the University of Heidelberg was seeking new asteroids on photographic plates when he found something unusual. It appeared to be an asteroid with a cometlike orbit that carried it among the inner planets. The vagrant asteroid had its orbit calculated and turned out to be an earth-grazer, possessed of an orbit that took it across and inside the orbit of the earth. This unusual object was designated 1932 HA and subsequently christened Apollo for the Greek sun god, because its orbit took it so near the sun. A few months later another Apollo-like asteroid turned up. This one was named Amor and was also an earth-grazer. Its perihelion took it to within one-tenth of an astronomical unit—less than ten million miles —of the earth's orbit. Ten million miles is barely a handbreadth as astronomical distances go, and it was something of a shock to see how closely our world was being "sideswiped" by rocks as big as towns.

Earth-grazers kept appearing. Soon astronomers had catalogued more than thirty of these cometlike asteroids and had classified them as Apollo-Amor objects, or Apollo objects for short (Fig. 7-1). Keeping track of them was difficult, because their orbits were unpredictable. There was no guarantee that a given Apollo object would show up again in the expected spot on its next turn around the sun, for Apollos,

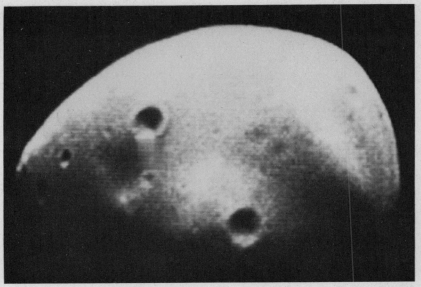

Figure 7-1. Apollo Objects. No closeup photos have ever been taken of Apollo objects ("extinct" comet nuclei), but most likely Apollo objects resemble the moons of Mars, one of which is shown here.

like comets, were constantly having their orbits perturbed by the gravitational fields of Jupiter and the inner planets. Apollo vanished shortly after its discovery; no one saw it again until 1973, more than forty years after it was first observed. Adonis, another earth-grazer spotted in 1936, disappeared in similar fashion and was rediscovered only in 1976.

Elusive as Apollo objects can be, astronomers have learned much about them from their fleeting visits. The largest known Apollo has a diameter of about six miles: this giant, known as 1978 SB, crosses not only the earth's orbit but that of Venus as well. Among the smallest Apollo objects is Ra-Shalom, a planetoid only about the size of a baseball diamond, that was discovered in 1978 and named to commemorate a recently negotiated peace between Israel and Egypt: hence the double name Ra (the ancient Egyptian sun god) and *Shalom* (Hebrew for "peace"). Apollo objects are a colorfully named bunch. The Aztec deity Quetzalcoatl has an Apollo object roughly half a mile wide named for him which was discovered in 1953. Sisyphus, the eternal rock-roller of mythology, has his own Apollo object, one of the biggest at a diameter of about three and a half miles. So do Daedalus, the father of technology, and Bacchus, the god of wine and festivity. Even Cerberus, the three-headed dog that was said to guard the gates of hell, has his name assigned to an Apollo object. These are only a few of the total population of Apollos. Estimates vary, but most astronomers think the Apollos number at least 200 and more likely about 750. That means there is one Apollo object for approximately every 530 asteroids.

Generally, Apollos have a specific gravity of about 3.5, only slightly more than that of a typical comet. Apollos are usually dark objects with a hue like that of potting soil, and their darkness combines with their tiny dimensions to make them very hard to spot, even when one knows where to look for them. All the same, Apollos reflect enough light to make spectroscopic studies possible, and recent observations of Apollo objects have shown that their surfaces are made up of the same materials found in carbonaceous chondrites.

Apollo objects bear many similarities to comets. Both categories of objects have eccentric orbits that take them close to the sun. They have comparable sizes, specific gravities, and chemistries (assuming, of course, that the chemical analysis of the fallout from the Tunguska event reflects accurately the composition of comet nuclei). About the only significant difference between Apollo objects and short-period comets is that comets have tails.

No comet, however, keeps its tail forever. After a few hundred revolutions, the comet burns out and ceases to release gas and dust on its approaches to the sun. It then becomes a small, dull chunk of rock circling the sun in a highly elliptical orbit that carries it into the vicinity of Mercury and Venus and out to the neighborhood of Mars, Jupiter, and the asteroids. In short, a burned-out comet will behave much as an Apollo object does.

Does this mean that the Apollos are extinct comets that have had their gaseous fires snuffed out? Öpik put forth that idea in the early 1950's, and today it is widely accepted. An "outgassed" comet, its volatiles driven off by solar heat until the nucleus no longer sports a tail, would look much like the known Apollo objects. Moreover, we can see comets —Encke's Comet among them—that appear to be losing steam and slipping into the Apollo category. In a few more decades or centuries Encke's Comet may cease to glow entirely, and become a bona fide Apollo object.

As noted earlier, Apollo objects and comets are both noted for complicated orbital mechanics; but some components of their orbits are fairly predictable, including their precession. Apollo orbits are known to precess, or swivel, around the sun at a more or less steady rate, under the influence of the planets' gravitational fields. An Apollo object's orbit will make one complete turn of 360 degrees around the sun in a few thousand years. Every five thousand years or so, orbital precession brings a typical Apollo object around into an orbit that intersects the earth's. The result may be either a near miss or an impact.

Near misses occur with surprising frequency. In 1937 the two-mile-wide Apollo object Hermes passed within eight hundred thousand miles of the earth, or approximately three times the distance from here to the moon. That passage was the astronomical equivalent of having a cigarette shot out of one's mouth in a circus show. Even closer passes have been recorded. Another and smaller Apollo object, perhaps the size of a boxcar, came within a few hundred thousand feet—not miles—of colliding with the earth in 1972. The planetoid plowed into the atmosphere over the western United States and flew north toward Canada, lighting up the night sky all over the Rockies and trailing a shock wave that reached the ground as an impressive sonic boom. Had the Apollo object's path carried it ever so slightly lower, the giant rock would have fallen somewhere in the Canadian province of Alberta and caused another Hiroshima on the prairie. The impact would have blasted out a crater hundreds of feet wide and subjected the surrounding area to a withering shower of heat and hard radiation.

Sheer luck saved the Canadians from disaster. At about sixty miles of altitude, the planetoid encountered the dense lower layers of the atmosphere over Montana and skipped across them, like a stone across a pond, before finally bouncing off the air and returning to space. Again in carnival terms, this near miss was like having the circus sharpshooter part one's hair with a bullet.

What happens when one of these interplanetary "bullets" doesn't miss? In that case the dead comet nucleus falls to earth as the Tunguska object did. The earth's face is scarred with hundreds of silent testimonials to its encounters with burned-out comet nuclei.

EIGHT

———— Comet Impacts

THE DISCOVERY OF CHUBB CRATER

One day in 1950, a Canadian prospector named Frederick Chubb was scanning Royal Canadian Air Force (RCAF) aerial photos of northern Canada. They showed a generally bleak landscape, sparsely covered with vegetation and dotted with narrow, elongated lakes. But one lake was different. It was almost perfectly round and stood out from the neighboring lakes like a bowling ball in a knife rack.

Chubb was intrigued. The round lake appeared to have a raised rim and looked much like a caldera, the bowllike formation sometimes found atop the ruins of extinct volcanoes. Was this a sign of a volcano in northern Canada? If so, then it might prove priceless, for diamonds are often found inside extinct volcanoes where the gems crystallized under intense heat and pressure while the volcanoes were still active. The magic word *diamond* in mind, Chubb went to see director V. B. Meen of the Royal Ontario Museum of Geology and Mineralogy in Toronto. Was this formation, Chubb asked, volcanic in origin?

No, said Meen. Yet the lake was something of equal interest to a geologist. Looking at the pattern of the lake, a central water-filled depression surrounded by an almost perfectly circular raised rim, Meen realized that this formation was probably an impact crater some ten thousand feet across, or twice the width of the famous Barringer meteorite crater in Arizona. The key word here was *probably*. There was still room for doubt about the origin of this spectacular formation. Only one thing would determine the truth, and that was a visit to the site, in far northern Quebec near the icy mouth of Hudson Bay. A newspaper publishing company agreed to have the expedition flown to the crater, and on July 17, 1950, the crater hunters took off from Toronto in an amphibious plane. Chubb and Meen were aboard with four other men: the pilot, an engineer, a reporter, and a photographer.

They flew northward across an increasingly bleak and unearthly landscape. Trees disappeared from view. So did all signs of animal life. At last they spotted the lake, or rather its raised rim, which from a distance looked like just another hill. On approach, they saw the lake itself, its surface ice-covered, even in July. The ice would have made landing on the lake hazardous, so the pilot set the plane down on another lake nearby.

After landing, Chubb and Meen set out on foot toward Chubb Crater, as they had named the round formation, two miles away. The trek was not easy, for the ground was covered with boulders and sloped uphill all the way to the crater. Along the way, Chubb and Meen had to climb over two ridges that would later help to establish the crater's origin.

At last the two men reached the edge of the crater and looked down into it. "We were so awed by what we saw," Meen wrote in a 1951 article on the expedition for *Scientific American,* "that I don't believe we spoke or even shook hands." He described the view before them:

Hundreds of feet below us lay a perfectly circular lake, cupped in a crater whose walls rose steeply in a slope of 45 degrees. No sound broke the stillness except the continuous grinding of ice on the water far below and the wind blowing across the crater rim. . . . We were looking down into what may well be the greatest crater of its kind anywhere in the world.

The dream of diamonds vanished quickly as Meen proved beyond any reasonable doubt that the crater had been made by a meteorite impact and not by a volcano. The ridges that Chubb and he had climbed on the way to the crater were circular ripples from the impact, like those of a pebble tossed into a pond, only much larger and frozen in the very rocks of the earth.

OTHER GIANT CRATERS

Chubb Crater was the second great meteorite scar, or astrobleme, to be positively identified on the earth's surface. The first was Arizona's Canyon Diablo crater near Winslow. Like Chubb Crater, the Arizona formation looks unimpressive from ground level. It resembles an ordinary gray mesa set in a reddish desert. Only when seen from above or from the rim does the crater take on its familiar scooped-out appearance. The Canyon Diablo crater is approximately fifty thousand years old. It owes its pristine state to the dry environment of the Arizona desert. There erosion by water is negligible, and so the crater remains much as it appeared just after it was created. Erosion has worn down the much older Chubb Crater slightly. The crater was knocked out long before the last glacial age and was ground down slowly as the heavy ice layers moved across it. The overlying ice protected the crater from attack by wind and water, however, so the net effect of the ice age was to preserve the crater more or less intact.

Though larger than the Barringer crater, Chubb Crater is not, as Meen imagined, the greatest astrobleme on earth. Since the Chubb-Meen expedition, photos made from aircraft and from satellites have revealed the traces of dozens of giant impact craters. Many of them are located on the Ca-

nadian Shield, the great expanse of primeval rock that
stretches between Canada's eastern and western mountains. Some of these craters are half the estimated age of the
earth. We can still identify them because they, like Chubb
Crater, were protected under thick ice for long periods and
thus avoided the destructive effects of erosion. The glaciers
wore down their rims slightly but preserved most of their
characteristic features.

Here are a few of Canada's more prominent astroblemes:

• Brent Crater, Ontario. All but invisible from ground
level, this two-mile-wide astrobleme is more than half a billion years old and is partially filled by a pair of lakes. Its
telltale circular shape was first observed on RCAF photos.

• Deep Bay Crater, Saskatchewan. Six miles across and
seven hundred feet deep, this crater is thought to be roughly
60 million years old. It was originally much deeper but has
been partially filled with debris deposited by glaciers.

• Gulf of St. Lawrence. The two-hundred-mile-wide gulf
between the St. Lawrence River and the Atlantic Ocean is
almost surely a drowned meteorite crater, for it bears all the
classic physical features of an impact structure: circular
shape, central depression, and raised rim. The coasts of New
Brunswick and Nova Scotia, from Miscou Point in the west
to Cape St. Lawrence in the east, trace an almost perfect arc
of a circle. Just across the Northumberland Strait, in the
gulf itself, Prince Edward Island follows the arc of a smaller
concentric circle. This pattern is difficult to explain by any
hypothesis except a planetoid impact. It appears, then, that
the Gulf of St. Lawrence was once a fiery gash in the earth's
surface, ripped open by the fall of an extinct comet nucleus
more than a mile in diameter.

• Hudson Bay. Although Hudson Bay has not been identified conclusively as an astrobleme, it does have features
that point to an extraterrestrial origin. For example, the
Nastapoka Island arc on the eastern shore of the bay is an
almost perfect semicircle, just like half of a meteorite crater.

• Lake Manicouagan, Quebec. Now occupied by a ring-shaped reservoir, this structure is the eroded ghost of a meteorite crater more than sixty-five miles in diameter and more than 200 million years old. Manicouagan's width is roughly equal to the distance from New York City to New Haven, Connecticut, or from San Francisco to Sacramento.

These "calling cards" of comets are not restricted to North America. Apollo objects have left impact craters all across the globe, from Australia to Arabia to Africa. One of the most impressive astroblemes is the famous Vredevoort Ring near Pretoria, South Africa. Sometimes spelled "Vredevort" or "Fredefort," the Vredevoort Ring is a faint but unmistakable remnant of what was once an impact crater as big as West Virginia. The Vredevoort Ring arose suddenly one day, perhaps a billion years ago, as an Apollo object one or two miles in diameter smashed into the earth from the southwest. The resulting fireball scorched the surface of the land for hundreds of miles around and released a blaze of light so brilliant that it could have been seen on the planet Mars. The explosion lifted up rock strata almost ten miles thick and overturned them as easily as you flip a page of this book. The whole earth shuddered as the fireball illuminated half a hemisphere as brightly as a giant magnesium flare. The explosion literally blew the top off the atmosphere over the impact site, and the shock wave that radiated out horizontally through the atmosphere struck with the force of a locomotive, sweeping away and pulverizing anything in its path. The Vredevoort impact had the same effect as the nuclear bomb that leveled Hiroshima in 1945, but many orders of magnitude worse.

AN APOLLO OBJECT'S IMPACT

Not even the most spectacular works of nuclear engineers can match an Apollo object's impact for sheer destructive power. To illustrate, let us envision another hypothetical comet like the one you rode in your imagination two chapters earlier. This particular comet is much older and virtually inactive. A short-period comet, it has made thousands of circuits of the sun, on each pass losing a small fraction of its volatiles, until now it has evaporated down into a mere drab lump of rock. This comet has been extinct for thousands of years. It ceased to glow in the nighttime sky long before the Pharaohs raised the pyramids.

If you stood on this comet's surface, you would no longer be surrounded by a gaseous coma, nor would you see the gray-and-white patches of ice on the comet's solid face. The frozen gases were volatilized long ago. All that remains is a gray-brown surface of porous silicate rock that resembles frozen mud.

The nucleus is roughly spherical and a little over three miles in diameter at its widest point. On one side is an impact crater several hundred feet wide, gouged out when the nucleus hit some stray bit of celestial debris, and then enlarged by the scouring action of escaping gases over the following centuries. On the other side of the nucleus is a great, rough fracture surface created when a piece broke off during a close approach to the sun. Everywhere are uneven holes "blown out" by jets of gas escaping from beneath the surface. It is a dark, drab, and desolate landscape, even more forbidding than the surface of the moon. Once a glowing ornament in the skies, the comet now looks like some huge cinder or a giant sponge cast in stone. It has become an Apollo object.

This Apollo object is not listed in the records of astronomers. Its dark surface and (by astronomical standards) tiny size have combined to keep it hidden from view. It has passed close by the earth on several occasions in the last few hundred years, but has never quite swum into the view of terrestrial stargazers. Even the giant orbiting telescopes lofted into space by the United States and the Soviet Union have failed to detect this nearby member of our solar system, partly because the telescopes were trained on more distant parts of the universe.

Very soon this ancient comet will be no more. Its orbit is carrying it on a collision course with the earth, and in a few hours it will annihilate itself in a huge fireball somewhere on the earth's surface. While we have a chance, then, let us look at the Apollo object carefully. It has an awe-inspiring set of statistics that will help us to visualize, later on, what it will do on impact.

For the sake of convenience, let us set the volume of the Apollo object at exactly thirty-five cubic kilometers. (We will use metric measurements here because they make calculations easier. Metrics will be translated into English equivalents as we go along.) That is about eight cubic miles. To find the nucleus's mass, all we have to do is multiply the volume times the specific gravity, or mass per unit volume. Here 2.0 is a reasonable figure, for it is the value estimated for Halley's Comet on its 1910 visit. So each unit volume of the Apollo object weighs twice as much as an equivalent volume of water. That specific gravity reading makes the extinct comet a featherweight compared to the earth, which has a far greater volume and an overall specific gravity of about 5.5.

Nonetheless, the comet has an impressive mass: roughly 70 billion metric tons, or 77 billion English tons. That works out to about 17 tons of Apollo object for every person on earth, or enough to fill a cube some 7 feet on a side.

Even a single ton of planetoid falling to earth will pack a heavy punch when it lands. A multibillion-ton mass will hit with much greater force. The question is, just how hard will the Apollo object hit? For the answer, we must calculate its kinetic energy—the physicist's expression for the energy carried by a body in motion. Don't worry; the calculation is simple, and it may give you a new respect for the laws of physics in action.

Kinetic energy is measured in metric units called ergs. One erg is a tiny amount of energy. Drop a nickel from waist level, and it will build up some hundred thousand ergs before hitting the floor. A man of average height and weight, walking at a brisk pace, carries roughly a billion ergs, while a typical car moving at turnpike speed has a kinetic energy measured in trillions of ergs. Small as they seem, however, ergs amount up fast because kinetic energy increases not merely with velocity, but as the square of the velocity. The kinetic energy equation looks like this:

$$K.E. = \tfrac{1}{2}\, m\, v^2$$

$K.E.$ here stands for kinetic energy, m for the mass of the object in grams, and v for its velocity in centimeters per second. Double the velocity of the object, and its kinetic energy increases by two times two, or fourfold. Triple the velocity, and kinetic energy increases nine times.

Now let us figure out the kinetic energy of our Apollo object. We have its mass and need a velocity reading. Twenty kilometers (about twelve miles) per second, relative to the earth, seems reasonable. Many meteoroids fall through the earth's atmosphere at about that speed, though velocities of a hundred kilometers per second and even faster have been recorded. (By contrast, a revolver bullet travels only about three-fourths of a kilometer per second as it leaves the gun's muzzle.)

Plugging the mass and velocity values into the equation, we get:

$$K.E. = \tfrac{1}{2}\,(7.0 \times 10^{16}\ \text{grams}) \times (2 \times 10^6\ \text{cm/sec})^2$$

The right side of the equation, when all the calculations are done, will give us the kinetic energy in ergs. The product of mass times velocity squared is:

$$K.E. = 1.4 \times 10^{29}\ \text{ergs}.$$

If that figure looks unimpressive, think of it as 14 with 28 zeros after it, or 140,000,000,000,000,000,000,000,000,000. That's a lot of energy, even measured in such tiny units as ergs. Then stop and realize that the nuclear explosion that leveled Hiroshima released only about 900,000,000,000, 000,000,000 ergs, or 9×10^{20}. Therefore our imaginary comet nucleus carries some 155 million times the energy released by the atomic bomb that destroyed Hiroshima in 1945.

That bomb was only about 15 kilotons in yield. In other words, its explosion released as much energy as setting off 15,000 tons of TNT. Today bomb yields are more often measured in megatons, each megaton the same as a million tons of TNT. A little simple arithmetic, then, reveals that the Apollo object's kinetic energy is equivalent to slightly more than 200,000 megatons. By comparison, a full-scale nuclear war between the United States and the Soviet Union would probably involve only about 10,000 or 20,000 megatons. So when the Apollo object hits the earth, the impact will be ten to twenty times more powerful than World War Three.

If even those numbers are hard to grasp, then conceive of the impact this way. It will amount to the same thing as exploding 200,000 one-megaton nuclear bombs. Each of those bombs can be put inside a cylinder about a foot wide. Now visualize, if you will, a row of one-megaton bombs—each one capable of demolishing Boston or St. Louis or San Diego— laid out, one bomb every seventy-five feet, all the way from the Brooklyn Bridge in New York City to the Golden Gate Bridge in San Francisco. That is how much explosive power the Apollo object will release when it hits the earth.

The extinct comet nucleus will strike in mid-Atlantic at 50 degrees north latitude and 42 degrees west longitude, almost equidistant from Newfoundland and Ireland. What follows is an account of the planetoid's last hour.

Countdown to Impact

Impact Minus Sixty Minutes. The Apollo object is 43,200 miles from its point of impact, coming in from just above the plane of the ecliptic. Its approach is unobserved. No radar on earth is aimed at this corner of the sky just now, and the dark surface of the nucleus makes it all but invisible against the backdrop of space. A keen skywatcher with a good telescope might notice an occasional occultation, or brief eclipse, of a star as the Apollo object passes between it and the earth, but such a coincidence would be remarkable, and the observer would probably dismiss it as a twinkling effect caused by turbulence in the air.

Impact Minus Forty-five Minutes. An observer on the planetoid would have a breathtaking view of the earth. The blue-and-white bulk of the planet would appear to grow bigger with each passing minute. Dusk has just fallen over the Eastern Seaboard of the United States, and a blaze of light delineates the northeast industrial corridor from Boston to Washington, D.C. The outline of Manhattan stands out boldly. So do the shores of Massachusetts Bay and Hampton Roads, their coastal cities sparkling in the twilight. The Midwest lies under clouds tinged red and violet by the sunset. Farther west, the green shores of California and the Pacific Northwest shade into the browns and whites of the snow-topped Rockies and Sierras and Cascade Ranges. The beige deserts of Arizona and New Mexico are virtually cloud-free, as is most of the Atlantic, where bright points of light gleam here and there in the darkness, marking the locations of planes in flight and ships at sea.

Impact Minus Thirty Minutes. Now the Apollo object is in what astronauts call the near-earth environment: only about a tenth as distant from the earth as the moon. As it sweeps past a high-altitude TV relay satellite, the Apollo object's bulk cuts off the satellite's signal for a moment. Technicians at ground stations are puzzled, but the signal returns as strong and clear as ever, and programming continues as normal. No one watching TV on earth notices anything out of the ordinary until a few minutes later, when the Apollo object will make its presence known in a more dramatic fashion.

Impact Minus Fifteen Minutes. The point of impact is now only about ten thousand miles away. Soon the Apollo object will encounter the outer fringes of the earth's atmosphere.

Impact Minus Five Minutes. Less than four thousand miles to go. A passenger on the planetoid would see the illuminated shores of Europe coming into view around the far edge of the planet: the brilliant sprawl of London, the lesser lights of Dublin and Belfast and Lisbon, the little sparkle of the naval facilities at Gibraltar.

Impact Minus Two Minutes. Now the Apollo object meets the first tenuous wisps of gas that mark the edge of the earth's atmosphere. This thin gaseous envelope has no effect at first on the huge rock. Yet even at this altitude, countless millions of air molecules strike the planetoid's face each second. Each molecule absorbs a bit of kinetic energy and turns it into thermal energy; and as the nucleus starts decelerating, it also begins to heat up—slowly at first, then rapidly.

Impact Minus One Minute. The air grows thicker every second as the nucleus passes over the Atlantic provinces of Canada. All at once air friction raises the nucleus's face to the point of incandescence. A bright yellow sheath of ionized gas surrounds the object and streams backward from it like a Valkyrie's hair. The leading edge of the dead comet's face glows red, then orange, then yellow, and finally dazzling white like the light of a magnesium flare. The tortured molecules on the planetoid's face are emitting more than visible light; they are also giving off radiation in the X-ray, radio, and television frequency bands. This sudden, fierce outburst of electromagnetic energy disrupts broadcasting over half the world. As viewers and listeners at home try to figure out what is happening, they see outside a light brighter than the sun at midday, moving in a stately manner across the sky. In cities from Boulder to Boston and Anchorage to Orlando, people look up to watch the giant meteor. The light is too intense to view directly. It casts dark shadows on the ground. Eerily and steadily, the shadows rotate as the meteor passes from west to east across the sky. The light is blinding but also curiously silent, like a gigantic starshell falling from beyond the moon. Moments later, the shock wave from the Apollo object's passage reaches the ground. The noise is not merely a roar or boom, but a sound beyond description.

Impact Minus Thirty Seconds. Less than a thousand miles from its impact site, the nucleus is dropping faster now. Air friction has robbed it of a trace of its momentum, and its path arcs downward slightly. The meteor lights up the skies from Moscow to Minnesota.

Impact Minus Ten Seconds. The meteor showers the area for a hundred miles on either side of its path with lethal doses of X-rays. No one has time to die of radiation sickness, however, for only seconds later . . .

Impact. In a fraction of a second, the energies of a dozen Armageddons are released at a single point in the North Atlantic.

NINE

Comets and Catastrophes

THE DEADLY WAVES

Imagine you are standing on an eastern beach, such as Fire Island or the Outer Banks of North Carolina, when the nucleus hits. Assuming you were not blinded by the glare from the planetoid's pass through the atmosphere and your sight survived the even brighter pulse of light released on impact, you would see the fireball, yellow and red and white and violet, rising up over the horizon like some great incandescent eye peering over the rim of the planet. The glare, like a hundred sunrises rolled into one, wipes out the stars and casts everything in the landscape—trees, grass, homes—into stark light and shadow. The most frightening thing about the fireball is its silence. There ought to be a roar, but none can be heard, for the noise will not reach the ocean's shores for some time to come; and when it does arrive, there may be no one left alive to hear it.

The fireball soars up through the atmosphere for ten, twenty, thirty miles, riding a convection current of its own making, to the very top of the earth's layer of air. All the way to Colorado and Saskatchewan, Iran and the Urals, a fierce yellow glow lights up the sky in the direction of the impact.

At last the fireball stops climbing. It has run out of atmosphere to rise in. The fireball cannot return downward, for still-rising convection currents continue to push it up from below. There is only one place to go—sideways—and so the fireball spreads out horizontally until it covers an area almost as big as Connecticut. Its volume is more than a million cubic miles, and the heart of the fireball is hotter than the surface of the sun.

Only a fraction of the total energy of the impact is turned into heat. Some of the rest is converted into hard radiation. A shower of gamma rays, capable of ripping through five feet of solid stone, irradiates the surrounding ocean, so that any ocean travelers not cooked alive by the thermal radiation of the fireball will be zapped with more than a thousand times the lethal dose of gamma radiation.

The most devastating of the strike's immediate effects, however, are the shock waves. They travel through air, sea, and land, and in their wake leave nothing but destruction. Earthquake waves ripple outward through the body of the earth from the impact site, as the whole world rings like a gong under the terrific force of the planetoid's fall. Because rock is a dense medium well suited to carrying shock waves over long distances, the vibrations will reach even the antipodes, the far side of the earth, with much of their initial punch intact. No place on earth will remain unshaken.

The pattern of vibrations will be complex, because the waves traveling through the rocks fall into two categories: S-waves and P-waves. Each is deadly, but in its own way. S-waves are surface waves. They spread outward from the point of impact as do waves created by a stone dropped into a pond. Earthquakes commonly produce S-waves, and eyewitnesses to earth tremors have actually seen S-waves undulating along the surface of the ground in the same manner as a ripple running down the length of a shaken carpet.

A typical S-wave is only a few inches in amplitude. (Amplitude means half the distance from crest to trough.) It can topple statues from their pedestals and shatter rigid brick or concrete buildings. The S-wave from an Apollo object's impact will have an amplitude measured in yards, not inches, and within a few hundred miles of the crater the earth's surface will rise and fall perhaps fifty feet as the wave sweeps along. Ripples in the rocks adjacent to the impact site will be still larger. You can see where similar S-waves have been frozen in place around impact craters here on earth. The ridges surrounding Chubb Crater in northern Canada, as we noted earlier, are actually shock waves preserved from the impact. Even more spectacular formations can be seen on the moon (Fig. 9-1). The mile-high Montes Appeninus mountain range is part of the "petrified" shock wave sent out through the lunar crust from the impact that created the Mare Imbrium millions of years ago. The moon's surface was uplifted thousands of feet in that wave's passing. The S-wave from our imaginary impact may not be that colossal, but it will carry enough power to raise whole mountain ranges and then drop them again as easily as you lift this book.

What will happen when that shock wave reaches the Atlantic shores? Think back to that shaken carpet mentioned earlier. When you shake a carpet, dust particles fly up from its surface. Much the same thing will happen to anything and anyone on the earth's surface as the S-wave sweeps by. Humans, houses, trees, livestock, and automobiles will all be tossed skyward like so many dust motes. Destruction will be total, or so near totality as to make no difference. No structure is built to take such stress. Neither are any organisms except perhaps the insects and the protozoa, which may be the only survivors of the S-wave.

If you could watch the S-wave's progress without being killed, then you would see the land undulate in a great S-shaped ripple as the shock front moved along. Buildings, roads, and other structures would appear to climb the face of the earth wave, disintegrating along the way. They would hang poised for a moment at the crest and then slide down the rear face of the wave, transformed into an unconsolidated sheet of rubble hidden under the cloud of dust stirred up by the S-wave's passage. This is only the initial S-wave.

Figure 9-1. The Moon. Our planet's satellite shows the scars left by colli-sions with Apollo objects. The lunar "seas," or "maria," visible in this photo as great dark patches on the moon's surface, were created when Apollo objects hit the moon and sent tides of molten rock oozing out over the landscape.

Others will follow but will add little to the massive destruc-tion caused by the first. The end result will be a level ex-panse of churned earth mixed with the debris of what used to be humans and their habitations.

If the S-waves were the only disturbances spawned by the impact, then its effect would still be literally world-shatter-ing. S-waves, however, are only one variety of the deadly waves. Their cousins the P-waves, short for pressure waves, are killers too. P-waves travel much deeper than S-waves. A

P-wave takes the low road, so to speak, and follows a short-cut through the heart of the planet, finally rumbling back to the surface somewhere hundreds or thousands of miles from its origin.

Tsunamis

So much for waves in the rocks. What about waves in the ocean? A comet nucleus hitting in the ocean will cause a tremendous tsunami, or seismic sea wave. The tsunami is one of the most destructive forces of nature, and one of the most mysterious as well. Humans have a long history of experience with tsunamis, especially in the Pacific Ocean basin, where a belt of intense earthquake and volcanic activity known as the Ring of Fire encircles the ocean and generates tsunamis frequently. So we already know what the tsunamis from a comet impact will do.

In deep water, the tsunami is no danger. The wave is barely perceptible as it passes a ship at sea, because the wave's length is so much greater than its height. A tsunami may be a hundred or even a thousand miles long, but only a few dozen feet high. Near shore, however, the length/height ratio changes. The wave starts to "feel bottom" in the shallow coastal waters and rises to form a colossal breaker that can sweep away anything within miles of the coast. Tsunamis have obliterated whole cities and even entire islands on occasion, and are capable of multiplying the destructive force of an earthquake or a volcanic eruption a thousandfold. Observe what happened following the Alaskan Good Friday earthquake of 1964. A sudden movement of rock along the Denlai Fault in southern Alaska set off a tremor and, with it, a tsunami that battered communities as far south as California. Thirty-foot breakers smashed the waterfront at Seward, Alaska, lifting freight cars and locomotives as easily as if they had been folding chairs. The fishing fleet at Kodiak was demolished. At one point on the southern Alaska shore, a fjord funneled the tsunami into a tiny bottleneck and so concentrated the wave's power that

water splashed more than a thousand feet (almost the
height of the Empire State Building in New York City or
the Sears Tower in Chicago) up the walls of the fjord. Visitors to the site later could tell exactly how high the water
had risen by noting the smashed and stripped trees on the
mountainsides. After traveling all the way down the coast
of Pacific Canada, Washington State, and Oregon, the
tsunami was still powerful enough to do serious damage
along the shore near Crescent City, California. Altogether
the wave caused more than a billion dollars in damage,
measured in 1964 dollars.

This is what the energies of one earthquake accomplished. Our hypothetical Apollo object will release far more
energy when it hits the earth, so its effects will be propor-
tionately greater. What is more, the planetoid will concentrate its energy release at one spot rather than spread its
force out over many square miles as the Good Friday quake
did. The pinpoint focus of the impact means that energy will
be transferred to the water much more effectively than in
the Alaskan quake, and will send a tsunami of titanic proportions radiating out at a velocity of hundreds of miles per
hour from the point of impact.

That tsunami will come ashore as a breaker the size of a
skyscraper on shores all around the Atlantic basin. In
theory, a breaker one mile high is possible. The actual
breaker would probably be considerably smaller, because
the continental shelf (the submerged portion of the continental land masses) would intercept the waves and make
them start to shoal before they reached the coast itself. Yet
even a comparatively small breaker, only five hundred feet
high or so, would inundate everything to a distance of many
miles from most shorelines: everything, rather, that was
left intact following the passage of the S- and P-waves. Anything that was not shaken down in the earth tremors will
then be drowned in the mammoth sea wave.

This tsunami is only the first of many. Others will follow,
as the earth heaves and shudders under the terrific blow of
the nucleus's impact and transfers the mechanical energy of

earthquakes to the waters. The seas will slosh back and forth in their basins for a long time to come, rushing over the land and then retreating, again and again, until the outlines of the continents as we know them have been altered forever.

Perhaps a few humans in protected places, far from the killer seas and lucky enough to survive the fury of the tortured earth, could endure these effects of a "superstrike" in mid-ocean. The survivors might even be able to replenish the world's population, provided that the earth tremors and tsunamis were the only major consequences of an Apollo object's fall. Unfortunately, the destruction will not stop here. The aftermath of a dead comet's collision with the earth will include yet another calamity, so far-reaching and enduring that it would further—and literally—darken the prospects for human survival.

Climatic Effects

Water will hurtle into the fiery throat of the crater, be vaporized, and go rocketing skyward as a column of steam and plasma, atoms stripped of their outer electrons by the unbelievable heat of the impact. As the tower of plasma and vapor rises, it enters chilly layers of the atmosphere and begins to cool. Then water condenses out of a vaporous state and forms tiny ice crystals, smaller than the finest snowflakes. Mixed in with the ice are salt crystals from evaporated seawater, and pulverized rock and sediment carried up from the ocean bottom. Together these fine particulates spread out through the upper air in the form of a slightly muddy ice cloud, more than a mile thick and thousands of miles wide. This cloud will have a catastrophic effect on world climate and consequently on all life on earth.

Suppose you could watch the spread of this cloud from some protected observation post on the European shore. It reaches Europe at dawn. As the sun rises in the east, the cloud moves in from the west and spreads a great dark stain across the sky. At the leading edge are wispy white ice clouds, tinted pink by the rising sun's rays; but behind those rosy tendrils looms a threatening mass of blackness, moving relentlessly toward you at ten miles' elevation. Far away on the horizon you can see brilliant white flashes of light from the titanic lightning bolts being generated in the turbulent air just around the comet's impact site. About the lightning bolts, the sky flares violet. Then the cloud is overhead and about to blot out the sun.

The sun is about a hand's width above the horizon when the cloud overtakes it. As the first feathery edges of the cloud drift across the disk, the sun starts turning red, because the finely divided solid material in the cloud is filtering out the shorter wavelengths of light near the blue and violet end of the spectrum, and allowing only the long red and orange wavelengths to penetrate. The sun fades to blood-red and finally to black as the cloud eclipses it completely. Darkness in the morning: that is the comet's lingering legacy.

Far overhead, the cloud is reflecting sunlight back out into space. Climate experts would say it is increasing the earth's albedo, the scientific term for the reflectivity of the planet's surface. Clouds are very effective reflectors, with albedos of 75 percent or more. That figure means they bounce three-fourths of incoming sunlight back into space and allow only 25 percent or less to get through. Since the world depends on sunlight to warm its surface, the planet will suddenly get much colder and less hospitable for life. Living things will suffer in other ways as well. Green plants rely on sunlight to make food, both for themselves and for the animals that feed on them. The plants will shut down in the darkness, and in a very short time, food will be hard to find. Millions of species are bound to perish.

Will any life be able to endure this global disaster? Perhaps, if the cloud is not too long-lasting. It is hard to tell how long the airborne particles will remain in the upper atmosphere, however, because once up there they are very difficult to remove. They are so fine that the slightest air currents suffice to keep them aloft. Even the steady bumping from gas molecules in the air, known to physicists as Brownian motion, can counteract gravity and prevent the dust and ice from falling back to earth. Moreover, much of the high-level cloud from the comet's impact will be suspended in the stratosphere, the zone above the "weather layer." Storms seldom penetrate into the stratosphere, so there is no rain or snow up there to wash the particles out of the air. Consequently the cloud may stay aloft for years or even decades before it finally dissipates.

We have no way of telling exactly how much this comet-made shroud may chill the earth, but we have some bench-mark data provided by volcanic eruptions that have cast whole cubic miles of ash into the upper air and had serious effects on climate. The Indonesian volcano Tambora erupted in 1815 and put out enough ash (thirty-six cubic miles, more than a hundred times greater than the 1980 output of Mount St. Helens) to virtually cancel summer the following year. New England and the eastern provinces of Canada had no warm season in 1816. Snow fell in Quebec in August. That year also saw the bitterest winter in then current memory, and went down in history as "the year without a summer" and "eighteen hundred and froze to death."

Tambora did this with only a few cubic miles of ash, most of which remained in the lower atmosphere. The volcano may actually have put no more than five or six cubic miles of particulates into the upper air. The impact of a comet nucleus—an Apollo object—will throw up a much greater cloud; the vaporized water alone will surely amount to a thousand cubic miles of salt and ice crystals, and the total may reach five thousand cubic miles. To help visualize that quantity, imagine excavating Chicago and its suburbs to a depth of about ten miles. Lift up the resulting chunk of rock, grind it into powder, and blast it into the air with the force of a million-megaton nuclear explosion. That is roughly what the nucleus's impact will do.

For at least a few years after the impact, the world will lie dark and cold under its high-level cloak of ice and dust. Even after the particles fall out of the air, it will be a long time before warm breezes blow again, because of something else the climate change will generate: ice and snow. Like clouds, frozen water has a high albedo. A snow field may reflect 90 percent or more of the sunlight falling on it. The bigger the snowfield, the more energy it returns to space; and it is safe to say that the global cooling following the comet impact will result in lots of snow. Sunlight reflected by the snow will be lost to space, lowering ground temperature still further, thus producing still more ice and snow . . . and so on, in a self-perpetuating cycle. Long after the particulates have fallen out of the upper air, the snowy ground cover will continue to keep the earth's albedo high and its temperatures low. In short, the earth will enter a new ice age.

Here of course we are extrapolating from known laws of physics and rather skimpy data. Humankind, so far as we know, has had the good fortune never to witness a catastrophe like the one described here. But species in previous ages of the world may not have been so lucky. If evidence from the fossil record is any indication, a similar disaster occurred roughly sixty-five million years ago and put an end to the dinosaurs. Indeed, you and all the rest of the human species appear to be here courtesy of a series of comet impacts that doomed the dinosaurs and gave our remote mammalian ancestors their chance for greatness.

DID COMETS KILL THE DINOSAURS?

The death of the dinosaurs has puzzled and intrigued scientists for more than two centuries, ever since fossil dinosaur bones and footprints were recognized as belonging to an extinct branch of the animal kingdom. At first dinosaurs were identified as victims of Noah's Flood: relics of God's original Creation that were left behind when Noah filled the Ark and drowned in the global deluge. This school of thought was called diluvianism. Its adherents cited the biblical Flood as the explanation for just about everything in the record of the rocks, including fossils. Diluvianism was a branch of "catastrophism," the view that the world is reshaped only by infrequent and dire events such as earthquakes, floods, and volcanic eruptions.

Catastrophism was popular through the eighteenth and early nineteenth centuries, but fell out of favor among scientists when two British geologists, James Hutton and Charles Lyell, proved in the mid-1800's that most change in nature occurs through the action of gradual processes working over vast lengths of time. The doctrine of slow but sure change is known as uniformitarianism and has prevailed in science for more than a century, for it presupposes that the processes at work on the natural world today are the same ones that operated millions and even billions of years ago. "The past is the key to the present" is the uniformitarian philosophy, and as a rule it has paid off handsomely for science. We are about to see, however, that some things, including the dinosaurs' extinction, are inexplicable in terms of the uniformitarian model but fit in neatly with the notion of comet-borne catastrophes.

The dinosaurs have received bad press in our time. They are commonly portrayed as hulking, witless beasts with brains no bigger than acorns. Some dinosaur species like the brontosaurs, whose brains accounted for only about half a millionth of their weight, did fit this image, but other dinosaurs were considerably brighter. *Deinonychus* ("terrible claw"), a hunter roughly comparable to the roadrunner of the American Southwest, was anything but slow and stupid. It ran down prey, grabbed them in taloned forearms, and gutted them with a slash from a long, razor-sharp claw on its hind foot. For this way of life *Deinonychus* needed a fast metabolism and a sophisticated brain, and clearly it had both. Many other dinosaur fossils demonstrate that dinosaurs were not primitive animals; on the contrary, they were superbly adapted to their environment and showed many features seen only in advanced mammals today. Dinosaurs were perhaps the most successful vertebrates (animals with backbones) that the world has ever seen. They dominated land, sea, and air, and filled ecological niches now occupied by species as diverse as dolphins, cats, and rhinos. The dinosaurs had everything going for them and by all rights should still be the dominant forms of life on our planet. Yet they are gone, almost as if they never existed. They vanished about 65 million years ago, in the space of about a million years, which is the blink of an eye as geological time is reckoned. Whatever killed the dinosaurs was no respecter of evolutionary progress. The smart ones died along with the mindless giants.

Many other forms of life died about the same time as the dinosaurs. The ammonites, squidlike sea creatures with elaborate curved shells, were abundant in the days of the dinosaurs but disappeared simultaneously with them. Marine limestone from the Mesozoic (the age of dinosaurs) is full of foraminiferans, tiny sea animals that lived in ornate glass shells similar in appearance to hot cross buns. Foraminiferans from the early Cenozoic (our current age of mammals) are smaller and look like minuscule golf balls. The division between these sections of geologic time is as sharp as a knife cut. It looks as if something just switched the Mesozoic off and the Cenozoic on. That something was most likely a comet impact that darkened and chilled their world, as in our earlier scenario.

Dinosaurs lived in a hotter world than ours. Paleoclimatologists (experts in ancient climates) have concluded that the world of 70 million B.C. was something like one big Puerto Rico, with climate so warm that tropical vegetation grew in the latitudes of present-day Connecticut, and corals lived in waters near what is now the ski country around Aspen. Accustomed as they were to a steamy environment, the dinosaurs would have had trouble coping with a world grown suddenly cooler following a comet's collision with the earth. They would have suffered serious thermal stress; and we see, in the fossils of the very last dinosaurs, hints that stress was indeed plaguing them. The fossilized eggs of some dinosaurs have been examined under X-ray analysis and revealed that the eggshells were too thick for the baby dinosaurs to open with their beaks. The eggs never hatched because the young were trapped inside. This mishap of reproduction indicates that stress was upsetting the parent dinosaurs' endocrine and other systems as the Mesozoic came to a close; and the evidence of marine fossils says that stress came from a rapid change of climate from hot to cool. The marine fossils of the early Cenozoic are overwhelmingly cool-water species—unlike the warm-water denizens of the late Mesozoic.

Evidence like this prompted a father-son team of physicists, Walter and Luis Alvarez of Berkeley, to propose that a planetoid impact destroyed the dinosaurs. The fall of an Apollo object, they pointed out in 1979, could have put enough particulate matter into the air to cause an abrupt global cooling and bring the dinosaurs to extinction.

An extraterrestrial doom! That idea seemed fantastic at first, but the junior and senior Alvarez backed their argument with convincing evidence from such sources as the Royal Society's enormous 1888 report on the climatic effects of Krakatoa's eruption.

Suspecting an Apollo object (that is, an extinct comet nucleus) as the agent of the dinosaurs' doom is one thing, but proving it is another matter entirely. Was there some pivotal bit of evidence, a "smoking gun," to link such an impact firmly with the extinction of the dinosaurs? There was, and the evidence came from chemistry of a thin layer of sedimentary rock.

In between two layers of limestone in Italy runs a rock stratum, only a few inches thick, that was laid down at the Mesozoic/Cenozoic boundary. Analysis showed that rock to be chock full of iridium, a silvery metal that is used to give color to paints and neon signs. Iridium is ordinarily rare in terrestrial rocks but relatively abundant in meteorites; and in this stratum the iridium concentration rose dramatically. Later study of other rocks of similar age, from Denmark and Colorado, turned up high concentrations of osmium, another metal that is scarce on our world but comparatively common in meteorites.

There was the smoking gun. Some major natural event appears to have injected large amounts of these rare elements into our planet's environment about the time the dinosaurs became extinct, and the most plausible mechanism for doing so is the impact of a comet or Apollo object.

The iridium- and osmium-rich rocks bring to mind a dramatic scenerio. A comet nucleus crashing into the earth some 65 million years ago pulverizes and vaporizes itself almost completely, and spreads its remains worldwide in the form of dust rich in rare elements. The dust settles on the waters, sinks to the sea floor, and is incorporated in sedimentary rock over the next few million years. Before the dust settles completely, however, it wipes out the dinosaurs and most of the other species that shared the Mesozoic world with them. Thus a comet writes the obituary of the dinosaurs, its peculiar chemistry announcing to their human successors what did in the Mesozoic giants and inaugurated the age of mammals.

The Alvarez hypothesis was persuasive, but many questions remained to be answered about the possible link between a comet impact and the mass extinctions of the Mesozoic. What about the impact structure, or crater? If an Apollo object did destroy the dinosaurs, it must have left a whopping big crater someplace. But where? Like detectives dogging a criminal's steps, geologists traced the iridium-rich rocks across Europe, and soon an interesting pattern emerged from their findings. The iridium concentration increased as one moved northwest from Italy toward Scandinavia and Greenland. It looked, then, as if the impact site must lie somewhere in that direction, for the iridium concentration would naturally be highest in the immediate vicinity of the impact. Was there any structure in the high-iridium zone that might have been knocked out by an Apollo object's fall? No clearly identifiable crater could be seen on the Atlantic floor, but right in the line of increasing iridium levels stood the volcanic island of Iceland.

Iceland perches on the Mid-Atlantic Ridge, an undersea mountain range that runs the entire length of the Atlantic Ocean from Arctic to Antarctic and includes several volcanic island groups such as the Azores. Iceland is so much bigger than the other islands on the ridge, however, that one has to suspect it had a different origin from the rest. Might Iceland have been built up of molten rock that oozed out of the earth after a comet nucleus blasted out a crater at that spot in the cold northern seas? That conclusion is tempting. Perhaps Iceland, then, formed the tombstone of the dinosaurs: a great mass of volcanic rock that rose from the midst of a comet's iridium-rich remains.

One Comet Impact . . . or Many?

Persuasive as all this evidence is, objections have been raised to the comet-impact account of the dinosaurs' end. If a comet really did cause the mass extinctions at the Mesozoic's close, then it would almost surely have wiped out the dinosaurs in a matter of days or weeks—a few months at the very most. Yet the dinosaurs did not vanish all at once. Some of them lingered on for hundreds of thousands of years after their cousins had vanished. Among these dawdlers were big plant-eaters whom one would expect to have perished immediately after an impact, when the cloud overhead killed off their food supply. Assuming a comet really did do in the dinosaurs, how did these species manage to last so long?

The fossil record does not quite fit the notion of a single impact wiping out all the dinosaurs at once. It does, however, support the idea of a series of impacts stretched out over a few hundred thousand years. Drs. Victor Clube and Bill Napier of the Royal Observatory at Edinburgh, Scotland, have suggested that toward the end of the Mesozoic, the earth passed through a "cloud" of comet fragments left behind when a giant comet, perhaps ten or twelve miles wide, broke apart millions of years earlier. Those fragments—each one capable of upsetting world climate—were traveling along in approximately the same orbit as the original comet that produced them. From time to time the precession of their orbit brought the comet debris close to the earth, and chunks of it started peppering our planet. The earth was then subjected to a steady bombardment of Apollo objects until orbital precession carried the comet fragments away. A prolonged onslaught of this kind would explain why the climatic upheaval of the late Mesozoic destroyed the dinosaurs, but not all at once. Instead, the comet debris picked them off gradually.

Thus comets have an important bearing on geology, paleontology, and other sciences that ordinarily are far removed from astronomy. Also, the extinction-by-comet scenario has significant implications for the whole philosophy of science, for it shows that the uniformitarians are not totally right, nor are the catastrophists totally wrong. The truth lies somewhere between these two extremes. Although the slow-and-steady principles of uniformitarianism govern the evolution of our world most of the time, there is still room in the picture for periodic catastrophes—including comet impacts —that clear the slate, so to speak, for new beginnings.

TEN

— Myths, Disease, and Comet Dust

HOW MYTHS RECORD NATURAL EVENTS

When comet nuclei fell to earth in prehuman times, they left a natural record of their visits in the form of impact craters, shatter cones, and other formations. Within historical times, there has been only one such collision—the Tunguska event—and scientists recorded that one in detail. Now what about the vast stretch of time in between these two periods, after the human species appeared in its modern form, but before humans had started keeping the systematic records that we define as history? Possibly comet impacts in human prehistory have been recorded and passed down to us in the less-than-scientific, but still revealing, body of "reporting" that we know as myths and legends.

Before a formal scientific literature arose, natural events were most likely to be recorded in personified form, as stories about the deeds of gods and goddesses. The tale of Persephone and Pluto, to cite a familiar example, seems to have its origins in the changes of the seasons, while the myth of Thor, the Norse thunder god with his colossal hammer, is thought to have arisen as a way to explain the fall of a large meteorite. On the way down, the meteorite would have set off thunder in the form of a sonic boom. Later, whoever found the meteorite itself lying in the middle of its impact crater could be excused for taking it to be a god's giant hammer. That god, the finder must have figured, hurls a hammer to make the thunder roll! Possibly this line of reasoning even gave the thunder god his name, for "Thor" is a fair transliteration of the noise of a sonic boom. In a comparable manner, comets may show up in various mythological guises.

Over the last few centuries, comets have been invoked to account for many myths and legends from antiquity. In 1694 Halley suggested that the "casual shock" of a comet had caused Noah's Flood by setting the seas in violent motion and sending them rushing over the earth. Possibly Halley was right. We have seen how tsunamis from the fall of an Apollo object at sea could inundate the land, and it may be significant that the account of the biblical Flood, in the book of Genesis, says the Flood began when "the fountains of the great deep [were] broken up" (Genesis 7:11). That passage suggests that the floodwaters did not fall as rain from the sky, as is commonly assumed, but instead rose up from the sea as a tsunami.

Whiston's *New Theory of the Earth*

Two years after Halley published his speculation about comets and the Noachian Flood, one of his colleagues, a mathematician named William Whiston, explored the same idea in greater detail, in a curious book entitled *A New Theory of the Earth.* Whiston, who succeeded Newton as professor of math at Cambridge, saw in comets a seemingly plausible way to explain Noah's Flood and a host of other phenomena as well. Whiston used a huge list of mathematical "proofs" and Scriptural references to back his assertions, and the result was one of the strangest works in the history of science.

All through Genesis, Whiston thought he saw references to comets. He suggested the primeval "chaos" out of which God created the earth was actually vapor in a comet's tail, and that this vapor condensed into the sun, moon, and planets. Here Whiston did not differ too greatly from the views of modern astronomers, who see comets and planets as products of condensation from a primordial nebula composed of dust and gas, perhaps four or five billion years ago. Very few scientists today, however, would agree with the rest of Whiston's ideas.

At first, he wrote, the earth had a perfectly circular orbit, a year of exactly 360 days, and a uniformly warm and pleasant climate. The earth did not yet rotate on its axis, so each day of the Creation was really 360 days long. (That lengthy "day" must have made belief in Genesis easier for rationalists who doubted that even God could fit so much activity into twenty-four hours.) Whiston also believed water vapor in the atmosphere had different optical properties then than it has now—a point to which we will return in a moment.

Everything was fine, according to Whiston, until Adam and Eve ate the forbidden apple in the Garden of Eden. At that moment God's Creation was spoiled, and the parents of the human species were expelled from Eden by an angel wielding a fiery sword. Though the angelic sword might be interpreted as an impression of a comet in the sky, that thought does not seem to have occurred to Whiston. As Adam and Eve made their way out of the garden, the great primordial comet mentioned earlier gave the earth a gravitational "kick" that set the world spinning on its axis.

There things stood for a few generations, until humankind grew so evil and corrupt that God could stand it no longer and decided to send a flood to cleanse the earth of people. In the Bible narrative, God gave the human race a second chance by warning Noah of the impending Flood and ordering him to build the Ark. Noah did as he was told and was safe aboard the Ark with his family when, on Friday, November 28, 2349 B.C. (Whiston had the date worked out that precisely), the Flood arrived: in Whiston's version, on the tail of another comet.

Whiston claimed that vapor in this second comet's tail condensed in the earth's atmosphere and produced a forty-day and forty-night downpour that covered the earth's en-

tire surface with water. The force of the deluge knocked the
earth out of its perfectly circular orbit and into an elliptical
one, thus adding five days to the year and bringing the calendar up to its present length. Whiston thought the comet
also slowed down the earth's rotation slightly, but he was
unsure whether it did so by magnetic effects or by the sheer
burden of water.

At last the floodwaters started draining away into the
earth's interior. The Ark ran aground on the mountains of
Ararat, and Noah and his family emerged from the Ark to
begin the job of repopulating the world. As the survivors
made a sacrifice to God to show their gratitude, a rainbow
appeared in the sky as a sign that God would never again
destroy humankind for its sins. There were no rainbows,
said Whiston, before Noah's time. God supposedly altered
the optical properties of water vapor solely to emphasize his
pledge to Noah.

Whiston's Biblical commentary may seem absurd today,
but his contemporaries took his flood-by-comet theory seriously, because biblical mythology was so much a part of
English intellectual life at this time that hardly anyone saw
fit to question the tales on which Whiston based his preposterous book. Even Isaac Newton seems to have accepted
Whiston's weird hypothesis as plausible. Ignatius Donnelly
carried on the Whiston tradition in his strange treatise
Ragnarok, and about half a century after Donnelly, a self-
taught cosmologist named Immanuel Velikovsky used a hypothetical comet to explain some famous Old Testament
stories. In so doing, he set off a controversy that still rages.

DR. VELIKOVSKY'S COMET

Born in Russia in 1895, Velikovsky attended a variety of
schools in Europe and the Middle East, and finally took an
M.D. degree at Moscow University. He spent some time in
Palestine as a general practitioner, then specialized as a

psychoanalyst. He moved to New York with his wife and daughters in 1939, and the following year he had an idea that was to make him famous (or notorious, depending on one's point of view) in scientific circles a decade later.

"It was in the spring of 1940 that I came upon the idea that in the days of the Exodus, as evident from many passages of the Scriptures, there occurred a great physical catastrophe," wrote Velikovsky. Exodus is the book of the Bible that recounts the Israelites' captivity in Egypt, the plagues sent by God on the Egyptians, the parting of the Red Sea, and the fall of manna from heaven to feed the hungry Israelites wandering in the desert. Not all Bible scholars believe these events really happened (it is thought that many of the Old Testament tales are intended as allegories rather than factual reporting), but Velikovsky assumed they did, and he spent the next nine years rummaging through the Columbia University library in search of evidence to back his literalist interpretation of the Scriptures. He found facts aplenty and eventually wrote them up in a book called *Worlds in Collision*.

Published in 1950, *Worlds in Collision* is the exciting story of a string of natural disasters that were recorded, Velikovsky claimed, as the miracles of Exodus. Velikovsky's tale is long and complicated, but the part of it dealing with comets may be summarized roughly as follows:

The planet Jupiter cast out a giant comet that passed near the earth on two occasions and finally became the planet Venus. The comet's first approach to the earth coincided with the Israelite Exodus in approximately 1500 B.C. and made the earth stop spinning on its axis. As the planet screeched to a halt, oceans spilled out of their beds, storms swept the globe, and mountains toppled. This tumult caused the Red Sea to part just as Moses and the children of Israel stood on the shore with Pharaoh and his army in hot pursuit. The waters withdrew precisely long enough for the Israelites to walk across the narrow stretch of sea bottom that separated them from the opposite shore. Then the sea returned, right on cue, to drown Pharaoh and his troops as they tried to pursue their former slaves across the exposed sea floor. Velikovsky thought the Red Sea really was red on that occasion. He proposed that the comet dropped from its tail a rain of reddish iron-bearing dust that turned the seas the color of blood.

Other things supposedly fell from the tail of the comet as well, including hydrocarbons (which Velikovsky imagined had seeped into the desert sands and become the source of modern-day Arab oil), and a sweet edible carbohydrate (manna) that precipitated out of the tail within our atmosphere and dropped to earth like a shower of breakfast cereal, saving the Israelites from starvation in the wilderness. Where that rain of oil came from is a mystery, since none has shown up in spectroscopic studies of comets. And the fall of manna? Velikovsky said it was made of starches, or carbohydrates, that were chemically the same as hydrocarbons. This was like saying bread is chemically equivalent to gasoline. Also, Martin Gardner points out, Velikovsky neglected to explain why the manna failed to fall, as Exodus tells us, on every seventh day.

Velikovsky brought back his wonder-working comet two months after the estimated date of the Red Sea crossing to provide the pyrotechnics for Moses' descent from Mount Sinai. The comet then retreated for about half a century (said Velikovsky) and put in a final appearance at the exact moment when Joshua, engaged in battle with the Canaanites, ordered the sun to stand still in the sky. Once more the comet brought the earth's rotation to a halt. The sun did appear to stand motionless overhead, and Joshua won the battle, aided by a fall of "hailstones" (meteorites) on the enemy army. Meanwhile another wave of catastrophes swept the suddenly arrested world, including the earthquake that destroyed Jericho. When at last the comet's business with Israel was done, the comet settled into an almost perfectly circular orbit as the planet Venus.

Where Velikovsky Erred

Thus Velikovsky explained the events of Exodus. His comet theory was of course preposterous, for a variety of reasons. First there is the difference in composition between comets and planets. Contrary to Velikovsky's view, Jupiter could not have cast out the comet, because comets are essentially big balls of water ice with a chemistry vastly different from a hydrogen-based gas giant like Jupiter.

Moreover, Velikovsky apparently assumed comets were sandbanks like those in the Lyttleton model: big orbiting swarms of individual particles that (he claimed) rained down on the foes of Israel as meteorites when the comet brushed against our world. We know now that the flying sandbank model was in error; and even if it were correct, one is hard pressed to see how such a relatively lightweight collection of bodies could have the terrific gravitation needed to jerk the earth about as Velikovsky imagined. Velikovsky's comet halting the rotation of the earth would have been roughly equivalent to a housefly lifting a man off the ground by his hair. At the same time, the earth's gravitation would have pulled apart the swarm of particles that made up Velikovsky's comet as soon as it came near the earth; in that case the comet would have disintegrated on its first approach and been unable to return and aid Moses at Sinai.

Then there is the gross dissimilarity between comets and the planet Venus. Nothing in the solar system resembles a comet less than Venus does. Venus is millions of times more massive than any known comet and has a surface hot enough to melt lead, which is an unusual condition for an icy object like a comet nucleus. Regardless, Velikovsky clung to his cometary theory of Venus's origin even while proclaiming that Venus was in "an incandescent state"!

What about Venus's orbit? If Venus had originated, within recent times, as a comet caroming around the solar system like some cosmic pinball, then one would expect Venus to have a highly eccentric, stretched-out orbit. It does not. Venus's orbit is an almost perfect circle, the smoothed-out condition of which indicates that the planet Venus has been rolling around the sun unperturbed for billions of years, not performing wild acrobatics as Velikovsky imagined.

Velikovsky was highly selective in the evidence he chose. Like Claramontius before him, he ignored any facts and opinions that contradicted his preconceived notions. Whenever the laws of physics and chemistry proved incompatible with his literal interpretation of the Bible, Velikovsky merely made up new laws of nature that would let his marvelous comet do whatever he wished it to. In effect, Velikovsky turned the scientific method upside down; instead of drawing conclusions from assembled data, he settled first on a conclusion and then started looking for any kind of evidence to support it.

From a nonscientific point of view, Velikovsky's ideas were appealing. They were colorful and—more important—gave an air of legitimacy to old Bible stories, so that America's biblical literalists, the modern heirs of William Whiston, could once again feel comfortable in their beliefs in the face of modern science. Well-known religious writer Fulton Oursler praised *Worlds in Collision,* and even some scientific journalists were swept away by Velikovsky's thrilling vision; the science editor of one New York daily newspaper hailed the book as "magnificent," while another journalist said that Velikovsky's gaggle of wild tales put him in the same category with such giants of science as Galileo and Newton.

Yet Velikovsky's comet hypothesis was so implausible from a scientific viewpoint that it probably never would have come to widespread notice had not an unusual chain of events catapulted it to prominence. Velikovsky's manuscript had been accepted for publication at one house when several scientists, outraged that such a book could find its way into print, put pressure on the publisher to drop *Worlds in Collision.* The publisher complied, but soon *Worlds in*

Collision was picked up by another publishing firm and is-sued despite all the scientific community could do to stop it.

This clumsy effort at censorship gave Velikovsky's work an aura of legitimacy that it could never have attained on its own. Velikovsky suddenly appeared to be a modern Gali-leo fighting against an oppressive orthodoxy, and so it was widely assumed that Velikovsky stood for some kind of rev-olutionary truth that the "scientific establishment" was trying desperately to conceal. Consequently Velikovsky has had a fiercely loyal coterie of defenders ever since the publi-cation of *Worlds in Collision* in 1950, despite the fact that his comet hypothesis has all the scientific validity of a Buck Rogers comic. Now and again some professional scientist will take time and trouble to refute Velikovsky's bizarre claims, but such criticism has made little impression on his defenders, for whom he will probably always be the arche-type of the persecuted genius.

Did Velikovsky really believe the story that he concocted? Apparently he did, for he stuck to it in spite of all criticism, from its publication to his death in 1979. A tall, impressive-looking man with white hair and a sharp nose, he looked like the popular image of the Old Testament figures he so admired, and he displayed (appropriately) Job-like patience when "orthodox" scientists, who put more faith in the known laws of physics than in the literal words of the Bible, reviled him as a crackpot and a pseudoscientist. "If I had not been psychoanalytically trained," he told one inter-viewer, "I would have had some harsh words to say to my critics."

NORSE LEGENDS AND COMETS

As the Velikovsky case demonstrates, myths and legends are not the best support for scientific inquiry where comets or anything else may be concerned. Yet some tales from an-tiquity do resemble the projected effects of a comet nucleus striking the earth, and one need not twist the laws of nature to mark the similarities.

Among the most intriguing legends in this regard is one that Velikovsky himself cited in *Worlds in Collision*. The story concerns the Lapp god Jubmel, Lord of Heaven, and his awful judgment on a sinful world. According to the legend, Jubmel looked upon the earth one day long ago and saw that humankind had become evil and corrupt. As God did in the biblical story of the Flood, Jubmel decided to wipe out the human species and start over. Jubmel descended from heaven in person to supervise the destruction of the world. The Lapps described him as a brilliantly luminous being wreathed in "fire serpents." No one could bear to look at him directly. That is not a bad description of the fireball from an Apollo object's impact.

As soon as the god set foot on earth, the ground began to shake. Great cracks opened in the soil, and men and women fell into them and were killed. One would expect something similar to happen as the S-waves from a comet impact tore up the landscape. But the earthquake was only the start of Jubmel's tantrum. Speaking directly to the doomed peoples of the earth, he promised to overturn the world (another reference to S-waves convulsing the earth's surface?), then made the rivers flow backward and sent the seas crashing over the land as a gigantic wall of water. The tsunami from a planetoid impact at sea would have the same effects. At last Jubmel put out the sun, whereupon the earth was covered in darkness and filled with the screams of the maimed and dying.

From start to finish, this legend recounts what one would expect to happen in the aftermath of a comet nucleus hitting the earth. Velikovsky invoked his hypothetical comet to explain these events, and maybe he was not too far from the truth. A collision between the earth and a comet nucleus, or even a small fragment of a nucleus, would accomplish exactly the kind and degree of destruction described in the tale of Jubmel.

Other Scandinavian myths and legends bear a strong resemblance to a report of a planetoid strike at sea. Perhaps the most famous of all northern myths is that of Ragnarok, the dark final day of the world. Ignatius Donnelly was not the only creative mind inspired by the Ragnarok story, by the way; German composer Richard Wagner used the same myth as the basis for his opera *Die Götterdämmerung,* or *The Twilight of the Gods.* The original story is told in the tenth-century Icelandic poem *Volsupa* and the thirteenth-century *Prose Edda.* Here is how Thomas Bulfinch summarized the Ragnarok tale:

> It was a firm belief of the northern nations that a time would come when all the visible creation . . . would be destroyed. . . . The earth itself will be frightened and begin to tremble, the sea leave its basin, the heavens tear asunder, and men perish in great numbers . . . The sun becomes dim, the earth sinks into the ocean, the stars fall from heaven, and time is no more.

The gods would descend from heaven and do battle with the giants, and the earth would be shaken and convulsed in the process. The sun would turn red and eventually go out as the great wolf Fenris devoured its rays one at a time, while from the sea a giant serpent would rise to spray venom over the earth. The noise and tumult would be indescribable and would cease only with the destruction of the world.

Again, if we strip away some of the religious imagery from the tale, then it sounds very much like a description of an Apollo object striking the earth in mid-ocean. The "gods descending from heaven" might well have been a falling comet nucleus or a cluster of nuclear fragments produced when an original, larger nucleus broke apart just prior to landing. A similar event happened millions of years ago in what is now Canada, where a falling Apollo object split into two pieces immediately before impact and blasted out a pair of adjacent craters that became the Clearwater Lake formation. The Vredevoort formation in South Africa also has some earmarks of a simultaneous multiple strike. A fragmenting comet nucleus, then, could easily have produced the illusion of a team of deities descending from heaven.

The reddened and then extinguished sun is a convincing detail, because it is exactly what one would expect to see as the plasma jet from the impact site cast a cloud of particulates into the upper air and blocked off incoming sunlight. The very name "Ragnarok" is suggestive of such an event, for "Ragnarok" appears to be a contraction of the Norse words *ragna rökkr,* meaning "day of darkness," a concise description of the worldwide gloom that a comet impact at sea would cause. Finally, a huge serpent rearing up from the sea is an acceptable image for the jet of steam and plasma that would shoot upward from the point of impact following a "superstrike" in mid-ocean.

COMETS AND BIBLICAL IMAGERY

If these stories tell of a comet nucleus striking the earth, then one ought to find other, similar tales in different parts of the world, for the effects of such a catastrophe would hardly be confined to northern Europe. We do find such tales in the writings of other civilizations, notably in the religious writings that are central to much of our own culture: the Bible. Scattered through Scripture are many passages that appear to duplicate the details of the Ragnarok myth and the projected effects of a comet nucleus's impact. Note the imagery Isaiah uses in the following passages:

Behold, the day of the Lord cometh, cruel both with wrath and fierce anger, to lay the land desolate; and he shall destroy the sinners thereof out of it. For the stars of heaven, and the constellations thereof, shall not give their light: the sun shall be darkened in his going forth, and the moon shall not cause her light to shine. And I will punish the world for their evil, and the wicked for their iniquity;

and I will cause the arrogancy of the proud to cease, and will lay low the haughtiness of the terrible. I will make a man more precious than fine gold; even a man than the golden wedge of Ophir. Therefore will I shake the heavens, and the earth shall remove out of her place, in the wrath of the Lord of hosts, and in the day of his fierce anger. [Isa. 13:9–13]

How art thou fallen from heaven, O Lucifer, son of the morning! how art thou cut down to the ground, which didst weaken the nations! [Isa. 14:12]

And it shall come to pass that he who fleeth from the noise of the fear shall fall into the pit; and he that cometh up out of the midst of the pit shall be taken in the snare: for the windows from on high are open, and the foundations of the earth do shake. The earth is utterly broken down, the earth is clean dissolved, the earth is moved exceedingly. The earth shall reel to and fro like a drunkard, and shall be removed like a cottage; and the transgression thereof shall be heavy upon it; and it shall fall, and not rise again. . . . Then the moon shall be confounded, and the sun ashamed, when the Lord of hosts shall reign in mount Zion, and in Jerusalem, and before his ancients, gloriously. [Isa. 24:18–20, 23]

Here Isaiah describes a catastrophe that extinguishes the light from the sun, moon, and stars; makes the land shake violently; and evidently kills so many people as to make survivors as rare as gold. (The "Ophir" mentioned here is the location of King Solomon's legendary gold mines.) The earth "shall fall, and not rise again": that is what happens in a landslide, and great landslides would be sure to accompany the S-waves that radiated outward from the point of an Apollo object's impact. As for the fireball, Isaiah may very well be describing it when he mentions that "the windows from on high are open." A window admits light into a room, so what could be more appropriate than to describe the burst of light from the fireball in terms of the opening of a heavenly window? Add to all this Isaiah's account of Lucifer

(the fallen angel whose name meant "light-bearer") dropping from the heavens to the ground—in other words, a
large and brilliant entity falling out of the sky—and there is
certainly reason to think that the prophet was drawing his
symbolism from memories of an Apollo object's collision
with the earth.

The New Testament both echoes and amplifies Isaiah's
imagery. Observe how Matthew described the cataclysms
that were expected to accompany the return of Christ:

> For as the lightning cometh out of the east, and shineth
> even unto the west; so shall the coming of the Son of man
> be . . . the sun shall be darkened, and the moon shall not
> give her light, and the stars shall fall from heaven, and
> the powers of the heavens shall be shaken. . . . [Matt.
> 25:27, 29]

Luke adds a few details:

> And there shall be signs in the sun, and in the moon,
> and in the stars; and upon the earth distress of nations,
> with perplexity; the sea and the waves roaring. . . .
> [Luke 21:25, 26]

This passage bears a considerable likeness to the myths of
Ragnarok and the judgment of Jubmel. Here again we see
objects falling from the heavens, great disturbances in the
seas, the sudden darkening of the sun and moon, and so
forth. Might these passages describe, in less than scientific
terms, the aftereffects of a comet nucleus's impact? It is
tempting to interpret them that way.

One may put the same interpretation on the words of
Peter, who in his second epistle describes Christ's promised
return as a sudden and unexpected calamity marked by the
fiery doom of the entire world:

> But the day of the Lord will come as a thief in the night:
> in which the heavens shall pass away with a great noise,

and the elements shall melt with fervent heat, the earth also and the works that are therein shall be burned up. [II Pet. 2:10]

Terrific noise, intense heat, the heavens obscured, and widespread destruction. Could anything sound more like the impact of an Apollo object? Yes, something could—at least, something in the Bible. Note how the apostle John recounts his vision in the book of Revelation:

The first angel sounded, and there followed hail and fire . . . and as it were a great mountain burning with fire was cast into the sea . . . and the third part of the sun was smitten, and the third part of the moon, and the third part of the stars; so as the third part of them was darkened, and the day shone not for a third part of it. . . . [Rev. 8:7–8, 12]

Biblical commentators have traditionally interpreted that "great mountain burning with fire" to be a volcano, for the Mediterranean basin where John lived is a hotbed of volcanic activity. The text says, however, that the fiery mountain was cast *into* the sea, whereas a volcano would more likely be described as rising *from* the waters. Was the "mountain" in this case actually a comet nucleus falling through the atmosphere in a shroud of blazing gases? Such an event matches John's vision. That reference to a "third part" of the sun's light being cut off is also interesting, because the cloud from a comet impact could reduce incoming sunlight by 33 percent or more. (It is more likely that the "third part" here was merely a figure of speech intended to stand for a big percentage of something.)

The more closely one reads John's account, the more it resembles a report of a comet colliding with the earth:

And the fifth angel sounded, and I saw a star fall from heaven unto the earth; and to him was given the key of the bottomless pit. And he opened the bottomless pit; and

there arose out of the pit, as the smoke of a great furnace;
and the sun and the air were darkened by reason of the
smoke of the pit. [Rev. 9:1–3]

Substitute "vapor" or "particulates" for "smoke" in that
passage, and it sounds almost exactly like the cloud raised
by an impact at sea.

A comet impact would be accompanied by mighty earth
tremors and tsunamis; and looking to the sixth chapter of
Revelation, we read:

And . . . lo, there was a great earthquake; and the sun
became black . . . and the moon became as blood; and the
stars of heaven fell unto the earth, even as a fig tree
casteth her untimely figs, when she is shaken [by] a
mighty wind. And the heaven departed as a scroll when it
is rolled together; and every mountain and island were
moved out of their places. [Rev. 6:12–14]

John repeats this detail about the missing islands a little
later: "And every island fled away, and the mountains were
not found." (Rev. 16:20) The seismic sea wave from a planetoid strike could submerge an island. Much of the British
Isles, for example, would surely be inundated by the tsunami from our hypothetical impact described earlier. The
wave might even break over the tops of low mountain
ranges; remember the performance of that tsunami in the
southern Alaskan fjord following the Good Friday earthquake of 1964. The wave might completely obliterate islands and mainland coasts composed of sand and other
unconsolidated sediments, for it would pack enough power
to pick up millions of tons of sand and gravel and carry them
away. In that case whole islands would literally be missing
after the tsunami's passage.

Perhaps, then, we can see terrifying visions of comet impacts in these myths and legends from our civilization's distant past; and possibly a comet's collision with our earth
ages ago gave rise to some of the myths by which we live.

DO COMETS CAUSE EPIDEMICS?

If British astrophysicist Sir Fred Hoyle is correct, then there may also be some basis in fact for the old superstition that comets bring pestilence and plague. Hoyle suggests that the seemingly barren celestial bodies beyond our earth may actually be teeming with extraterrestrial life; we have just been seeking it in the wrong places. Rather than searching the other planets, says Hoyle, we ought to look for life on comets. More ominously, he has also proposed that bits of life from comets have been drifting down from outer space for ages and setting off epidemics—so that perhaps there really is a link between comets and disease, as the old superstition has warned all along.

Contagion from space is only one in a long string of intriguing ideas that Hoyle has put forward. His hypotheses often run contrary to conventional wisdom in science; and while they have often been overturned, the argument over Hoyle's ideas has forced many scientists to reexamine their own convictions, with highly productive results. Perhaps the most famous work of Hoyle's was his steady-state model of the universe. Unlike many other astronomers who believed that the universe was created in a single initial explosion (the "Big Bang") billions of years ago, Hoyle envisioned a basically static universe that had no origin and no end, that had always existed and always would, for its matter was supposedly being created and destroyed at equal rates, a few particles at a time. These were the main points of Hoyle's steady-state model, though it touched on many other topics, such as the "forging" of heavier elements out of hydrogen and helium in the hearts of stars.

Steady-state cosmology was debated hotly during the 1960's and had been largely rejected by the 1970's. Experiments turned up no evidence of matter being created or destroyed as Hoyle's model required. More dramatic was the discovery of a soft, steady hiss of radio noise from all over the sky. This noise was the slowly fading "echo" of the Big Bang. All the same, however, the steady-state model served a good purpose, for it explained perfectly how stars cook up, so to speak, the elements that make up our planet and our bodies. So when Hoyle speaks, other astronomers listen, for they know they will at least hear a challenging idea.

Nothing would seem more challenging, on first hearing, than the suggestion that comets actually do drop pestilence on the world. How could an iceball millions of miles away give rise to disease-causing microbes, much less scatter them across the face of the earth? Hoyle came up with a possible explanation that made the thought of diseases from space seem a bit less farfetched.

His argument may be summarized roughly as follows. The chemicals of life are not unique to our world. They are found in abundance all through the solar system and even in the space between the stars. Carbon, the atom on which all organic chemistry is based, has turned up in meteorites, asteroids, and the spectra of comets. Quite a few organic molecules such as cyanogen appear in comet tails. Mix those molecules together in liquid water, and you have a primordial "soup" of the kind in which the first living things are believed to have arisen billions of years ago. Such a "soup" is exactly what Hoyle has suggested we might find inside comet nuclei!

How could liquid water exist inside a comet's frozen nucleus? Perhaps easily, given a little heat from the decay of radioactive atoms in the nucleus, and an insulating layer of ice and dust to guard the liquid from the chill of deep space. In this protected environment, the chemicals in solution would be free to combine with one another in pools of water until they happened to come up with the basis of life as we know it: the nucleic acids. These acids are easy to synthesize out of raw, prebiotic chemicals in a laboratory, so nature might be able to produce them too, in a natural test tube like the one Hoyle envisioned within a comet nucleus. From nucleic acids it is only a short step to the simplest known form of life, the virus. Described by one biologist as "a bit of bad news wrapped in protein," a virus is basically nothing more than some DNA inside a protein sheath. The virus attacks a living cell by injecting its own viral DNA into the cell. This invasive DNA then takes over the cell's reproductive system and orders it to turn out copies of the virus. The cell obeys until it fills with viruses and ruptures, releasing the viruses to go off and assault still more cells.

Thus viruses cause diseases in humans, from polio to smallpox. When you have influenza, your lungs and windpipe feel sore because the cells in their linings are being attacked and "blown apart" by viruses; and you stand a good chance of getting the flu every few years, because a new and dangerous strain of flu virus appears every seven years on the average and starts making its way from person to person through the population. The result is an epidemic, or a pandemic if the disease should spread worldwide.

The most famous flu pandemic was the Spanish influenza of 1918. It killed more persons than World War I. Known in its time as "the plague of the Spanish lady," the 1918 flu left no land untouched. It also left behind a memory even more ghastly than that of the war, and the frenzied merriment of the Jazz Age, starting in the early 1920's, was probably a reaction to the twin horrors of the global conflict and the awful disease that followed it. Histories of the pandemic state that it arose in the Iberian peninsula and was spread overseas by troopships and civilian vessels. This is the conventional explanation. Hoyle and his colleague Chandra Wickramasinghe of Cardiff University in Wales, however, saw flaws in the traditional wisdom.

The virus, they pointed out, seemed to spread faster than the slow-moving transportation of the early 1900's would allow. Influenza broke out almost simultaneously in America and India, even though the sea route to India from Europe was much longer than to the United States. Indeed, the flu advanced at the speed of modern jet aircraft, at a time when passenger air service was nonexistent. If the virus was transmitted from person to person by physical contact, as the traditional theory went, then how do we account for the pandemic's lightning advance in an age of sluggish transportation?

Also, the flu struck some individuals and communities but left others in the same vicinity untouched. It was nothing unusual for one town to suffer casualties from the flu while an adjacent town was practically flu-free, despite frequent travel between the two communities. There would have been plenty of chances to spread the disease if the traditional model of contagion were right. So why didn't the flu spread?

Hoyle and Wickramasinghe looked for answers in the history of epidemiology, the science of epidemic disease. What if, they wondered, the flu virus had dropped from a comet and been transported by air?

Airborne disease was a popular concept in epidemiology during the eighteenth and early nineteenth centuries. Malaria, for example, got its name because Italian doctors thought the "bad air" *(mal aria)* over swamplands caused the sickness. When Louis Pasteur, Walter Reed, and other researchers established that many supposedly airborne ailments were in fact caused by microorganisms spread from person to person either directly (say, by shaking hands) or indirectly (through the bite of an infected flea or mosquito), the "bad air" explanation went out of fashion. To Hoyle and Wickramasinghe, however, the airborne-disease model seemed to tie in neatly with their thoughts about comets dropping microbes.

Imagine, the scientists said, a comet infected with pathogenic microorganisms (germs, if you prefer) passing across the earth's orbit and spewing them out in its tail. The result is a cloud of microbes in the earth's path. As the earth passes through the cloud, some of the microbes are swept up in the planet's atmosphere and make their way down to the ground. They do not fall in equal density everywhere, but are swirled around by atmospheric turbulence so that they settle in a patchy pattern: lots of germs here, very few over there.

So far, so good. Nothing in this hypothesis flatly contra-
dicts any known facts of science. A serious hypothesis, however, ought to do more than merely agree with what is already known. It should also be useful for making
predictions. This is the acid test of any hypothesis or theory,
and Hoyle and Wickramasinghe knew it. So they made a
prediction. Future flu outbreaks would be spotty, they said,
as one would expect if flu viruses were dropping down
through the air from space.

Spotty outbreaks are exactly what Hoyle and Wickramasinghe observed in a subsequent flu season. The scientists
studied the distribution of flu cases at schools in Britain and
reported highly patchy distribution patterns. Some dormi-
tories had numerous cases of influenza, while others only a
few yards away were practically free of the flu. Was this
proof that comet-borne flu viruses were falling from the
sky? Not necessarily; it was also possible that resistance to
the flu, rather than the flu itself, had been distributed in a
patchy pattern, by the laws of chance rather than by wind
and weather.

Some individuals are highly resistant to given strains of
flu. Other individuals are more susceptible to the viruses.
There is no guarantee that the "immunes" and the "suscep-
tibles" will be mixed together in equal numbers in any
given area. In fact, it is much more likely that one group
will predominate over the other in any certain locality, and
a patchy distribution pattern will develop. Statisticians call
this a clumping phenomenon. The clumping has a mathematical explanation but can be illustrated much more easily with the aid of a simple demonstration.

Go to a pet supply store and buy two boxes of colored
aquarium gravel, one white and one red. Take them home
and empty their contents into a glass jar, then cover the jar
and shake it up and down to mix the contents. You might
expect to find a totally regular distribution of colors: red,
white, red, white, just like the pattern on a chessboard. But
you don't. Red predominates in some places, and white in
others. No matter how long you shake the jar, you will prob-

ably never obtain an even, homogeneous mix of colors. There will always be clumps of red and white. Now imagine that each of those bits of gravel represents a human. Reds are susceptible to flu. Whites are resistant. It is easy to understand, after performing this experiment, how the patchy distribution pattern observed in the flu studies might have arisen. So there is no need to invoke comets to account for epidemic disease. More earthbound mechanisms seem adequate.

Yet the Hoyle-Wickramasinghe hypothesis does bear an uncanny similarity to some baffling cases in epidemiology. There are instances in history—not many, but enough to pose a significant puzzle—where a mysterious disease has appeared only once or twice, caused widespread illness, and then vanished, never to recur. One example is the "sweat," a frightening malady that struck Britain repeatedly, starting in 1485. The "sweat" was characterized by its sudden onset, rapid spread, and symptom of copious perspiration. In most cases the victim died from dehydration within hours after the symptoms first appeared. The sweat was unlike any other disease described before or since. It would terrorize Britain for a few months, then vanish, only to return a few years later for another brief but deadly visit. At last the sweat disappeared in 1551, apparently for good. Had it been the work of a native terrestrial microbe, one would expect the disease to have stayed around longer and recurred more often; but those were its only known appearances. If Hoyle and Wickramasinghe are correct, then the mystery of the sweat is easy to solve. The earth passed through a cloud of microbes left behind by a passing comet. The precession of the cloud's orbit then carried it away from the earth, and the sweat thereafter spared the world.

Improbable? Perhaps. It is a little hard to imagine, for instance, how microbes evolving on a distant comet could develop a "taste" for humans here on earth. On the other hand, many scientists very recently would have applied the word *improbable* to quasars, black holes, the giant volcanoes of Mars, and the underground sulfur ocean of Jupiter's moon Io. So we may wish to file the Hoyle-Wickramasinghe hypothesis in the "unproven" category until more conclusive data come in. Note also how comets are playing an important role in science ordinarily far removed from astronomy—in this case, epidemiology and the biology of microorganisms.

THE HUNT FOR COMET DUST

Microbes may not be dropping on us from passing comets, but other bits of comets do appear to be drifting down to earth, in the form of tiny dust particles spewed out in the comets' tails. Fred Whipple predicted in the 1950's that comet dust could fall unscathed through the atmosphere to earth because the individual particles are so tiny. Though the particles would enter the upper layers of the atmosphere at a relative velocity of several miles per second, the bits of dust would not burn up as meteorites would, because the dust particles, being practically massless compared to meteorites, would not have enough kinetic energy to immolate themselves. They would simply be swept up and would slowly make their way down through the air to the surface. Depending on the size of a given dust particle, it might take half a year to fall the sixty or seventy miles from the upper air to earth. As best we can tell, this is exactly what happens. Comet dust trickling slowly down through the air adds about ten thousand tons every year to the earth's mass. (This accumulation of dust is neatly balanced by the loss of terrestrial dust to space; an approximately equal amount of fine dust is carried from the earth's surface into outer space by air currents and molecular motion, and as a result the earth leaves a dust trail behind it as it revolves around the sun. Our planet's dusty "tail" can be seen sometimes in the evening sky as the *Gegenschein,* or "counterglow," a roughly triangular patch of faint light on the horizon that shows where the dust is reflecting sunlight.)

The infall of comet dust is extremely light by ordinary standards. It amounts to only about one or two particles per square yard per day, on the average. Yet it is a subject of intense interest to astronomers, for the comet dust dropping on us from space offers an opportunity to gather pieces of a comet and bring them into a laboratory for study.

Even if the comet dust contains nothing so exotic as alien microorganisms, it may still prove to be a scientific bonanza, for it could tell us much about the early history of our solar system. Remember that comets, if our current ideas about their origins are accurate, are relics from the distant past. They were formed hundreds of millions of years ago and therefore represent the primeval material of our solar system, the original stuff from which the sun and planets were formed. None of that material has survived here in the inner solar system, which has been reworked extensively since the birth of our system and thus retains no sign of what the "newborn" solar system was like. If we can lay hands on even a few minuscule chunks of cometary material, they will open a window on the distant past and allow us to clear up many mysteries about our little corner of the galaxy.

But collecting that material is not easy. The particles of comet dust at high altitudes are so few and far between that special collection techniques are required. About fifteen years ago United States researchers began using balloons to carry comet-dust collectors high into the atmosphere. One of these gadgets was something like a vacuum cleaner. It sucked in large quantities of air and filtered particulates out of them. The machine was known as the Vacuum Monster in honor of a cartoon creature in the Beatles' film *Yellow Submarine* that sucked in everything around it. The Vacuum Monster was powered by a rocket engine that burned hydrazine fuel, a good choice because it can be burned in the absence of oxygen, which is hard to come by in the rarefied upper air.

Air roared in through the Vacuum Monster's intake at more than three hundred miles per hour. The airstream passed through a special filter that was about as wide as an oatmeal box and traversed by sticky glass rods that captured any particles they encountered. When its fuel was exhausted, the Vacuum Monster dropped back to earth by parachute. Scientists then collected particles from the filter and set about analyzing them. Optical analysis was not very useful. Under an ordinary light microscope it was hard to tell an extraterrestrial bit of dust from grit off a sidewalk. Fortunately, the scientists had a better tool for the job of identifying the particles—a scanning electron microscope. The electron microscope bombards each sample with a powerful beam of electrons. This process generates X rays, the frequency of which is characteristic for each different substance that is zapped with the electron beam. So by noting the frequency of the X-ray output, the dust hunters can tell whether they have a possible bit of comet dust or something more mundane, like volcanic ash. Much of what the Vacuum Monster gathered was uninteresting, such as bits of paint from high-flying aircraft or tiny spherules of titanium oxide from the exhaust of spacebound rockets. Now and then, however, a probable bit of comet would turn up. It was easy to spot: the X-ray frequency revealed that this particular sample had a high carbon content like that of a carbonaceous chondrite. That chemical signature pointed to a cometary origin. Also significant was the fact that particles like this one had optical spectra very much like those of comet tails. That evidence added still more weight to the hypothesis that the Vacuum Monster had sucked up comet dust.

The Vacuum Monster met a tragic end. After a promising start (it gathered one probable piece of comet dust on its first flight and several more on its second), the Monster was plagued by gremlins. Its third flight was cut short when the balloon sprang a leak. The fourth flight was the Monster's last. Its parachute failed to open, and it plummeted twenty-five miles to its demise.

Now the Monster's job is handled by two converted American U-2 spy planes. The U-2 is a sleek twin-engined jet aircraft built in the 1950's to spy on the Soviet Union. Exceedingly lightweight (only about eight tons) and designed to fly thousands of miles without refueling, the U-2 has extremely long and slender wings that carry it into the thin upper atmosphere at altitudes of five miles or more. The U-2 made headlines in 1950 when one of the planes was shot down over Russia by a ground-to-air missile; the pilot was captured, and the incident seriously embarrassed the United States while wrecking plans for an upcoming United States–Soviet summit meeting. Nowadays the U-2 has more peaceful duties. Its reconnaissance chores have been taken over by camera-carrying satellites, and the U-2 now flies on missions for NASA to sweep the upper air for comet dust. The U-2's ability to fly high comes in handy on these missions, for the jet can rise far above the "filthy" lower air where pollutants would make the search for comet dust all but impossible. It can also fly much longer missions than the Vacuum Monster could, and thus increases the chance of snaring comet dust. Instead of passing air through a filter, the U-2 carries a set of adhesive-coated glass plates about the size of index cards in special airtight canisters at the tips of its wings, and sticks the plates out into the airstream at high altitude to gather anything that might cling to them. A U-2 on a given flight may pick up one promising particle per hour. Then the plates are retracted, the U-2 returns to its base in northern California, and the samples are analyzed.

The suspected comet particles do not look impressive. They show up under the electron microscope as lumpy agglomerations of smaller particles. They resemble clods of soil, or bread crumbs, or a collection of sea sponges. They also are thought to be much more revealing than meteorites of the chemistry of our solar system as a whole. Meteorites are unusual simply because they were able to survive their fall through the atmosphere. They appear to be unusually tough as interplanetary matter goes. So they are not a representative sample of matter in between the worlds. The carbon-rich but fragile stuff of the dust particles is probably much more indicative of what matter between the stars and planets, the original material that coalesced to build up the stars and planets, is like. Thus the study of this alleged comet dust is already giving us a better idea of what conditions prevailed at the origin of the solar system, and what the chemistry of the universe is like on a large scale.

The samples gathered by the U-2 may fill an important gap in comet studies during the years ahead. Though the United States has no plans to send space probes to investigate Halley's Comet during its 1985 approach, the Soviets, the Japanese, and the European Space Agency (ESA) are planning unmanned missions to the comet. Those probes and their goals will be discussed in Chapter 12.

COMETS IN OUR FUTURE

We can only learn so much about comets by studying them from earth. We could learn much more if humans could live on comets—and perhaps one day we will. Physicist Freeman Dyson, in his memoir *Disturbing the Universe,* foresees a day when comet nuclei will teem with living things, namely humans and their allied flora and fauna.

"Comets," Dyson writes, "not planets, may be the major potential habitat of life in the universe." Dyson points out several relevant facts about comets. They have abundant water, carbon dioxide, and other materials needed to support life as we know it. Comet nuclei have a total surface area thousands of times greater than that of the earth and in theory could support huge human populations. Finally, Dyson thinks genetic engineering may make it possible to create trees capable of growing on the cold, airless surface of comets. Plants could be equipped with airtight skins to keep in moisture. Self-contained greenhouses could make optimum use of sunlight when the trees are more than one astronomical unit from the sun. Such alterations might be tricky but not impossible, and a comet seeded with such plants could be transformed into a little living world, a hospitable home for humans in the bleak void of deep space. Passengers on the comet would live among the root systems of the trees and would return, in a sense, to the arboreal habitat of our ancient ancestors in Africa. Long ago we left the trees, and conceivably we will return to them one day as residents of tiny "comet states" circling the sun. Should comets turn out to be fit for settlement, they may, Dyson suggests, help speed the spread of humankind throughout the cosmos. Perhaps we can hop from comet to comet on journeys between the stars, much as the heroine leaped across the ice floes in nineteenth-century melodramas.

Any such voyages, however, are far in the future. For at least a few decades we will have to be satisfied to study comets at a distance, using spacecraft and earth-based telescopes. Fortunately for amateur astronomers, comet watching —unlike other branches of astronomy such as stellar physics—is still an open field for the nonprofessional, and a well-trained amateur astronomer working at home can contribute as much to our knowledge of comets as a Ph.D. at a large observatory can. Amateur comet watchers will play an important role in the observation of Halley's Comet during its 1985 return, and this is a good point for a brief discussion of the upcoming Halley watch.

ELEVEN

How to Watch for Halley's Comet

OBSERVING HALLEY'S COMET

Of all the comets in the sky
There's none like Comet Halley.
We see it with the naked eye,
And periodically.

—Sir Harold Spencer Jones

The 1985 to 1986 return of Halley's Comet has been heavily publicized, so it will probably be the most closely observed object in the skies around the turn of the year. Though professional astronomers of course have the best instruments for comet watching, the amateur astronomer can discern much about the comet. Equipment for watching Halley's Comet need not be expensive or elaborate; even a pair of binoculars can yield splendid views under the right observing conditions. This chapter is not intended to substitute for a comprehensive manual for comet watchers, but the suggestions offered here may help backyard astronomers prepare for the comet's return.

First, a word of warning: Do not be misled by exaggerated newspaper stories into thinking that the upcoming appearance of Halley's Comet will be as dazzling as, say, Donati's Comet in the past century. Halley's Comet will not turn night into day, nor will it stand out in the sky like a search-light beam. As comet experts Brian Marsden of Harvard University and Robert Roosen of NASA's Joint Observatory for Cometary Research stated in a 1975 article for *Sky and Telescope* magazine: "if you were disappointed by Comet Kohoutek in early 1974, don't have high hopes for a fine display from Comet Halley."

Several things will combine to make Halley's Comet something less than the spectacle of the century, at least when seen from the Northern Hemisphere:

1. Orbital Geometry. Halley's Comet will approach from the south. As a result the best viewing will be in the Southern Hemisphere, and comet watchers north of the equator will have less opportunity to watch the comet. From the latitude of New York City, observers will see the comet rise only about thirty degrees above the horizon. The view will improve as one goes south, but even in the tropical corners of the United States the comet will not rise more than forty degrees. This leads to the next problem facing United States and Canadian comet watchers.

2. Air Pollution. Many parts of North America—even supposedly pollution-free places like Arizona and Colorado—suffer from air pollution that can seriously affect comet watching. Since the comet will be low in the sky, much of its light will be cut off by smoke, dust, and smog. Better seeing conditions will prevail in the Southern Hemisphere, where the comet will appear more nearly overhead and will be less obscured by air pollution.

3. Light Pollution. This is another major problem in industrialized countries like Canada and the United States. Comet seekers may find their quarry lost in the blaze of light from large cities, or street lights, or the moon. "How spectacular the comet will eventually appear depends to a considerable extent on how serious the problem of light pollution has become by 1986," write Marsden and Roosen.

4. Weather. The comet will be closest to the sun when cloudy winter weather prevails in much of the Northern

Hemisphere. Consequently seeing conditions may be poor over much of the northern half of the world during the comet's visit. Even when the sky is clear, chilly temperatures may make outdoor observations uncomfortable, if not impractical. By contrast, the Southern Hemisphere will be enjoying summer weather at the time, and comet watchers there will find their work much easier.

Prospects for seeing the comet from North America, then, will depend to a large extent on the vagaries of weather and the proximity of bright light sources. Where are the best places from which to watch the comet? There is no easy answer to that question. No one can accurately give the reader a list of specific locations—such as Stowe, Vermont, or Bend, Oregon—with directions on how to get there, and promise optimum viewing conditions. The site may be covered with clouds or made inaccessible by winter road conditions at the moment the amateur astronomer cares to visit it.

So instead of listing specific sites for viewing the comet, let us offer a few guidelines for selecting a site.

How to Choose an Observation Site

1. Avoid light pollution as much as possible. This usually means seeking out a rural location. Observers in Boston, for example, may find the Berkshires an attractive location because of their distance from the city's bright lights. Even a trip of twenty miles to the western suburbs will improve viewing conditions greatly. You are the best judge of your local situation. Whether you travel five miles or a hundred miles is up to you. Residents of urban areas, however, should not give up all hope of seeing the comet. Remember that Japanese comet watchers (see Chapter 2) have an enviable record of discoveries even though their country is one of the world's most densely settled and polluted lands, in terms of both light and air pollution; and even in brightly lighted New York City, Halley's Comet was plainly visible on its last appearance in 1910. Your choice of locations may depend largely on how much detail you want to observe. If

simply seeing the comet is your only goal, then the city may
be fine for you; but for more exacting observations you will
probably have to remove yourself from town.

2. Avoid places with badly polluted air. Here again, rural
areas are the best bet, but even spots with allegedly pure air
may have enough pollution to hinder viewing, because of
pollutants wafted in from other parts of the country. Smoke
from Ohio Valley factories, for example, clouds the air and
reduces visibility over New England, a thousand miles
away. One possible countermeasure to air pollution is to
find some high elevation from which to watch the comet.
The higher up you go, the less pollution will stand between
the comet and you. The Rocky Mountains are full of such locations, as are the Sierras, the Appalachians, the Cascades,
and the Laurentides in Canada. Sites like these have another advantage: there are no buildings around to block
your view. High elevations also tend to be inhospitable in
winter, however, so ordinary caution is advised when pick-
ing a mountain or plateau for comet viewing. It makes no
sense to drive three hundred miles to Mount Wherever if
weather conditions stand to make the place inaccessible or
hazardous for traveling.

3. Read weather reports. Your results will of course be
best when the weather cooperates, as Edmond Halley discovered during his surveys of the southern skies (see Chapter 4). To ascertain the best times and places for comet
watching, consult your local forecasts—not the newspaper
forecasts, which represent an average of conditions over a
wide area and may not reflect conditions where you are, but
rather, the detailed local forecasts given by your local
weather bureau office or by a private weather forecasting
service. There may be a tremendous difference in weather
within a space of a few miles. (In San Francisco, for example, viewing may be good from the mountains just east of
the city but poor in the city itself, because of the fogs that
roll in off the bay and ocean.)

Choosing a Telescope

Simply finding a good spot to set up a telescope is not enough. You also need a good telescope or other viewing apparatus. If you are intrigued enough by Halley's Comet to go a considerable distance in search of good viewing conditions, then most likely you are interested in astronomy as a hobby and already have a telescope. If you have no telescope and wish to acquire one, it is easy to do, but you need a little background information about telescopes and their use before buying one. The telescope that works fine for one amateur astronomer may be inappropriate for another stargazer, for a variety of reasons.

Telescopes generally fall into one of two categories: reflectors and refractors. The refractor is the kind most people think of when they hear the word *telescope*. A refracting telescope consists of a long tube with glass lenses mounted at either end and is no different in its basic design from the old-time mariner's spyglass. The refracting telescope has served astronomers well for centuries and will do the same for you. It is compact and easily portable and is the favorite of many amateur astronomers.

The second kind of telescope is the reflector. Invented by Sir Isaac Newton, the reflecting telescope uses a mirror to gather light and focus it on another, smaller, mirror that is viewed through the eyepiece. Reflectors have some advantages over refractors. A reflector delivers images with less distortion, because reflectors have no lenses, which can "twist" an image out of shape and spoil viewing if they possess even a slight irregularity. Each kind of telescope has its defenders, and which one you choose is up to you. Used reflectors and refractors are ordinarily available at modest prices. When selecting a telescope you may want to ask the advice of other amateur astronomers in your area, for they may be able to steer you toward bargains. Do not judge a telescope by its outward appearance, because rust spots on the barrel do not affect viewing, and a wobbly mount can be

fixed with a few turns of a screwdriver. Much more important is the quality of the optical equipment inside the telescope. Check a telescope out carefully before buying it. If at
all possible, arrange to look through it under the conditions
in which you plan to use it, that is, searching the night sky
for faint objects.

For best results a telescope should have a wide aperture,
the diameter of the opening at the far end of the telescope,
with respect to its focal length, the distance between that
lens and the eyepiece through which you look. Divide the focal length by the aperture width, and you get the focal ratio.
For example, if your telescope is 20 inches long and has an
aperture width of 2 inches, then the focal ratio is 20/2 or 10.
The focal ratio is commonly written with an *F* prefix: in this
case, F/10. The greater the focal ratio, the more detail a telescope can provide. If you wished to see the fine structure of
lunar craters, for example, you would want a telescope with
a very long focal ratio. Comet watchers are looking for faint
objects in the skies, however, and therefore want the aperture to be wide so as to admit as much light as possible. So
the best telescopes for comet observations should have a relatively low focal ratio, on the order of perhaps F/3 or F/4.

When you hear talk of the *power* of telescopes, it is easy to
misunderstand the meaning of the word. Power in this case
is the ratio between the focal length of the eyepiece and the
focal length of the objective, meaning the rest of the telescope. If you have a telescope with a focal length of 30 inches
and an eyepiece with a focal length of one inch, then your
telescope has a power of 30/1, or 30. That is about the highest power you will need for comet viewing, and you can do
well on considerably less. Many of the famous comet hunters of the eighteenth and nineteenth centuries had great
success with telescopes of much lower power.

If you have never looked through a telescope eyepiece before, you may be surprised to see that the telescope delivers
an upside-down image of whatever celestial object you use it
to view. The lenses or mirrors invert the image when refracting or reflecting light. This is why many maps of bodies

like the moon are printed with the north pole at the bottom of the page and the south pole at the top. At first this inversion effect may be confusing, but you will get used to it soon. Fortunately for comet watchers, comets may be observed just as well upside down as right side up.

Where to Look for Halley's Comet

Even the best telescope is useless, however, unless you know where to look for the comet. Locating it is not merely a matter of pointing your telescope at a given constellation and watching the comet spring into view. You will probably have to do some concentrated hunting for Halley's Comet, at least before it nears the sun and develops a prominent tail. Detailed instructions for locating it are beyond the scope of this chapter. They would occupy most of a book—and in fact there is just such a book. *The Comet Halley Handbook,* by Donald K. Yeomans (NASA/Jet Propulsion Laboratories, 1981), is available from the U.S. Government Printing Office in Washington, D.C. Another invaluable book for comet watchers is *The International Halley Watch Amateur Observers' Manual for Scientific Comet Studies* by Stephen Edberg of the International Halley Watch (IHW). Details are available from International Halley Watch, Jet Propulsion Laboratory, Mailstop T-1166, 4800 Oak Grove Drive, Pasadena, CA 91109. The IHW was started to coordinate data gathered by comet watchers all over the globe, and will provide, upon request, an information sheet that tells amateur astronomers how to prepare for observations of Halley's Comet.

Amateur astronomers will play a major role in scientific observation of Halley's Comet on this visit, for several reasons. Amateurs now have extremely sophisticated equipment available to them and can carry out observations that earlier in this century were possible only for major observatories; therefore the professional astronomers can rely on the accuracy of a skilled amateur's observations. Then there is the time element to consider: many professional astronomers will be watching other phenomena in the skies and will have no time to watch the comet. Also, as *Sky and Telescope* pointed out in a 1983 article, the amateur comet watchers, unlike the professionals, "are not limited by the rather inflexible observing schedules imposed at large observatories." The magazine added that "amateurs are more evenly distributed around the globe and thus can keep an almost constant vigil. Finally, they just plain outnumber professionals. When all is said and done, amateurs will certainly accumulate more total hours observing the comet than will professionals."

As you watch for Halley's Comet, several measures will make the job easier for you:

• Don't start peering through the telescope immediately after you set it up outdoors. You probably won't see much of anything because your eyes will still be adjusted to bright indoor light. Wait about twenty or thirty minutes before you actually begin observations. By that time your eyes will be accustomed to the dark, and viewing will be much easier.

• Keep your sky chart handy. The sky chart is essential, even to the experienced amateur astronomer, because it shows precisely where everything is in the sky. The sky chart is especially important to comet watchers because it designates the star clusters and other faint, fuzzy celestial objects that might be mistaken for comets by an inexperienced observer. Better circle those noncomets in red to make sure you notice them while making observations. How can you follow the chart in the darkness? Easy. Just keep a tiny penlight in your pocket and shine it on the chart as needed. That way you can follow the chart without seriously impairing your "dark-adjusted" vision.

• Be patient. As mentioned earlier, the comet may not leap out at you the instant you first look through the eyepiece. You will probably have to do some searching. Since the telescope's field of view is limited to a small portion of the sky, you will have to scan the heavens in a very methodical pattern to make sure you find what you are looking for.

Many successful amateur astronomers prefer a series of sweeps, starting low in the sky and proceeding upward. Imagine you are reading this page backward, starting in the lower right-hand corner and proceeding line by line to the upper left. That is the kind of motion your search ought to follow. Do it slowly and carefully, and if something is there to be seen, you will probably see it.

• Be comfortable. That may be difficult if you are observing on a winter night when temperatures are hovering near freezing, but at least try for some measure of comfort while watching for the comet. That will free your mind to concentrate on observing. Some amateur astronomers choose to stand while looking through the telescope, while others prefer to sit. It's up to you.

Once you find the comet, what should you watch for? Halley's Comet offers plenty to see. This comet has done just about everything a comet can do, except break apart. The handbooks mentioned earlier will provide detailed suggestions for observations, but here are a few in summary:

• Astrometry. This is the five-dollar word for determining just where things are positioned in the sky. Astrometry is very important in comet studies because it is impossible to work out a comet's orbit successfully without very accurate position measurements. Astrometric techniques are a subject beyond the scope of this chapter, but they are not difficult to learn and can be carried out by amateurs.

• Spectroscopy. Today amateur astronomers have much better spectroscopic equipment than was available to the professional astronomer in the nineteenth century. For a reasonable sum you may buy a spectroscope ready-made, or if you are mechanically inclined you can build one yourself, using prisms or diffraction gratings (finely ruled transparent sheets of plastic that break up light into its component spectrum). Details are available from astronomical supply companies or from your local businesses that sell high-quality optical products.

• Meteors. Here professional astronomers are relying on amateurs to contribute the vast majority of observations. Note any "shooting stars" seen in the course of your comet watching.

• Photography. Astronomical photography is a demanding art. Time exposures are necessary for most shots, and for them the telescope must be kept trained precisely on the comet or whatever else is being photographed. "Clock drives" are used to swing the telescope slowly across the sky in time with the rotation of the earth and are available to the amateur at a modest price from astronomical supply companies or may be bought used at a considerable discount. Cameras are easy to mount on telescopes and may be arranged to take wide-, moderate-, or narrow-angle photographs. Wide-angle shots will show the entire comet from head to tail, while moderate- and narrow-angle photos are better suited for detailed studies of the tail and coma. If you take high-resolution, narrow-angle pictures of the comet, your photographs may be of great value to astronomy, for they may reveal details of the comet that have never been caught on film before.

TWELVE

— Scientific Studies of
—— Halley's Comet

STUDIES IN PROGRESS

The scientific study of Halley's Comet on its upcoming visit has already begun. The comet's faint image was detected on October 16, 1982, at Mount Palomar observatory in southern California. As the comet approaches the sun, large telescopes all over the world will of course be trained upon it, as they were on Halley's last visit in 1910. This time, however, astronomers will have tools that were not available to the stargazers of the early twentieth century, namely spacecraft. The Soviet Union, Japan, and the European Space Agency (ESA) are preparing to send unmanned probes to Halley's Comet in an effort to take measurements at close range.

The United States planned for a while to send a probe to Halley's Comet, but two factors combined to kill that project. One was a penny-pinching attitude in Washington. The federal government has spent the past few years cutting expenditures on scientific research in general and space research in particular, and the United States decided it could not afford an unmanned mission to the comet, despite the fact that the probe would have cost only the equivalent of one penny per taxpayer per day for less than a single year. The second factor was the set of priorities that the National

Aeronautics and Space Administration (NASA) uses in planning and scheduling space missions. NASA has long been committed to exploring the planets—the moon, Mars, Venus, etc.—to the exclusion of other celestial bodies such as comets. Therefore a comet mission, even to such a famous comet as Halley, did not have top priority with NASA, and a probe was not authorized.

In the late 1970's NASA proposed a Solar Ion Propulsion (SIP) probe to be sent to Halley's Comet in collaboration with ESA. This probe would have run on solar energy and, instead of relying on conventional rockets for propulsion, spewed out a stream of ions—charged atoms—to provide gentle but steady thrust. The SIP probe would have been able to "sneak up on" Halley's Comet, perform observations at leisure, and possibly even land on the nucleus and retrieve a sample of it for study.

The SIP project had much to recommend it, but the U.S. government decided against funding SIP research and development, and so the effort had to be abandoned. (This incident is said to be one reason why ESA decided to exclude the United States from participation in its comparable project: Europeans feared the United States would back out at the last minute to save a few pennies and would thus scuttle the whole undertaking.)

There is much to be learned from a mission to Halley's Comet. Close-range observations of the comet are expected to tell us much about the early history of our solar system, for comets, as mentioned earlier, are thought to represent the primordial stuff from which our system was made. So missions to Halley's Comet will (in effect) serve as "time machines" that will give us a glimpse of conditions that prevailed in our star's neighborhood hundreds of millions of years ago. These missions will be supplemented by observations from earth and from the United States Pioneer spacecraft in orbit around the planet Venus. The spacecraft will be reoriented to point its instruments at Halley's Comet. Pioneer is expected to provide astronomers with detailed views of the comet at its closest approach to the sun. The comet will be invisible from earth at that time because of the sun's intervening bulk, but the Venus spacecraft will be in a position to observe Halley's Comet as it reaches perihelion.

The ESA probe, named Giotto in honor of the Italian Renaissance artist who depicted the Star of Bethlehem as a comet in his famous painting "The Adoration of the Magi," is perhaps the most interesting of the upcoming studies of Halley's Comet, for the probe is intended to zip through its tail within about six hundred miles of the nucleus. This passage will be hazardous (it has been compared to a kamikaze attack in World War II), for the spacecraft will traverse the tail at the staggering velocity of some forty to forty-five miles per second relative to the comet. At that speed the probe could easily be destroyed if it should encounter any sizable bits of dust or grit from the nucleus. At that relative velocity a particle the size of a grain of beach sand would hit the spacecraft with the destructive force of an armor-piercing bullet. For that reason ESA is equipping the probe with specially designed armor. Even if the shielding saves the spacecraft from destruction in a collision with a particle, however, the mission might still end in disaster. The impact might tilt the spacecraft so that its antenna was no longer pointed toward the earth, and the probe could not announce its findings to the scientists who sent it up. If the antenna gets shifted even two degrees off target, its signals will miss receivers on earth entirely and the mission will be a loss.

However, assuming all goes well, the ESA probe and the Japanese and Soviet vehicles are expected to clear up many gaps in our understanding of the phenomenon of comets. For example:

• Does Halley's Comet have a solid "dirty snowball" nucleus, as the Whipple model claims? All the evidence we have indicates the Whipple model is correct, but the existence of the nucleus is still a matter of surmise and has yet to be proven. It is unlikely that probes will be able to photograph the nucleus itself on this visit of Halley's Comet (for that task a relatively slow-moving spacecraft like the SIP probe would be required), but radar imaging—the construction of "pictures" from patterns of reflected radar signals—may succeed in giving astronomers solid proof that a "dirty snowball" nucleus is present.

• What is the tail made up of? Our present knowledge of the tail's composition is based largely on spectroscopic studies made from earth and from satellites. Probes to Halley's Comet will provide an opportunity to expand that knowledge greatly, for the spacecraft will be able to sample the tail stuff directly, using instruments such as a neutral gas mass spectrometer, which identifies atoms and molecules that do not have electrical charges on them, and an ion mass spectrometer, which does the same for charged molecules and atoms. These instruments should return a bonanza of information, for they are expected to encounter one hundred to five hundred atoms and molecules per second as the spacecraft sweeps across the comet's tail. The data returned to earth may give us a better understanding of the physical and chemical processes at work in the comet: for example, which large molecules are being broken apart by solar radiation, and what the products of this decomposition are. There are many other uncertainties to be cleared up as well, such as the proportion of dust to gas in the tail. That information could tell astronomers much about the composition of the nucleus.

• How does the comet interact with the solar wind? For the first time, astronomers and physicists will be able to dip directly into a comet's electromagnetic environment and map the pattern of magnetic fields around it. The resulting data ought to show the accuracy or inaccuracy of our presently accepted models of comets and their interaction with the "wind" from the sun.

Scientists are also expecting the unexpected from the upcoming missions to Halley's Comet. Every time a space probe succeeds in its mission, we here on earth are rewarded with a bundle of surprises. No one expected to find giant volcanoes on Mars, or braided rings around Saturn—but they showed up on pictures returned to earth by United States space vehicles and gave us a better understanding of the other worlds in our solar system. Undoubtedly probes to Halley's Comet will reveal things we could never have guessed, and those revelations may completely change the way we look at the universe around us.

Glossary

ACCELERATION. Increase in velocity over time. *See* Velocity.

ACCRETION AXIS. In the Lyttleton model, the imaginary line, parallel to the direction of the sun's movement, along which particles coalesced to form comets.

ALBEDO. How much light a surface reflects; in more conventional terminology, the "lightness" or "darkness" of an object.

AMPLITUDE. The distance between the crest and the trough of a wave.

ANTIPODES. The point on the earth's surface exactly opposite from any other given point; in everyday terms, the other side of the world.

ANTITAIL. The sunward-pointing "spike" of gas and dust sometimes seen projecting from a comet's head.

APHELION. The point in a comet's orbit when it is farthest from the sun.

APOLLO OBJECTS. A class of celestial objects that appear to be extinct comet nuclei and have eccentric orbits that carry them near or across the orbit of the earth; sometimes called merely Apollos.

ASTEROIDS. A population of several thousand planetesimals in between the orbits of Jupiter and Mars. *See* Planetesimals.

ASTROBLEME. The impact formation, usually a crater, left by a planetoid that falls to earth. *See* Planetoid.

ASTRONOMICAL UNIT. The median distance between the sun and the earth; approximately 93 million miles or 130 million kilometers.

BASE LINE. The distance between two sightings taken during a parallax measurement. *See* Parallax.

CALDERA. A bowllike volcanic formation similar in appearance to an impact crater.

CARBOHYDRATES. Generally speaking, sugars and starches.

CARBONACEOUS CHONDRITE. A category of lightweight, crumbly meteorites. *See* Meteorites.

CATASTROPHISM. The school of scientific thought in which most or all change in nature is attributed to infrequent and catastrophic events.

CENOZOIC. Present age of life on earth; the "age of mammals."

CLATHRATE. Arranged in a lattice.

COMA. The roughly spherical cloud of gas and dust immediately surrounding the nucleus of a comet.

COMET. One of a population of small, icy bodies in orbit around the sun; in limited usage, either the nucleus of a comet or the coma, tail, or any other feature or phenomenon associated with a comet. *See* Nucleus, Coma, Tail.

CONJUNCTION. The meeting of two celestial objects at the same point in the sky.

CONVECTION. The vertical movement of heated fluids through a cooler surrounding medium; in more common usage, warm air rising.

CORONA. The extremely hot halo of gas immediately surrounding the sun.

COSMOLOGY. The study of the universe as a whole, its origin, its operation, and its possible future.

CYANOGEN. A toxic compound found in comets in very low concentrations.

DECELERATION. Decrease in velocity over time. *See* Velocity.

DILUVIANISM. The school of scientific thought, now discredited, that held the biblical Flood to be responsible for the major features of the globe.

DISCONNECTION EVENT. The loss of a comet's tail at a sector boundary. *See* Sector Boundary.

ECCENTRIC. Elongated, when used to describe an orbit.

ELECTRON. A negatively charged subatomic particle.

ELLIPSE. The geometric shape that describes the orbits of comets and planets; a flattened circle with two foci, or focal points, located on its long axis, instead of a single focus, as in the case of a circle.

EPICYCLE. A theoretical device once used by astronomers to explain retrograde motion of planets. *See* Retrograde Motion.

EPIDEMIOLOGY. The science of epidemic disease.

ERG. A very small metric unit of energy.

FORAMINIFERANS. Small aquatic animals with perforated glassy shells.

GEOCENTRIC. Earth-centered.

GRAVITATION. The tendency of any two masses to draw closer to each other; an intrinsic property of all matter.

GRAVITY. *See* Gravitation.

GRAVITY WELL. A mathematical simulation of an object's gravitational field.

HELIOCENTRIC. Sun-centered.

HYDRATE. A water-bearing chemical compound.

HYDROCARBONS. In very general terms, petroleum compounds.

HYDROXYL. A single oxygen atom combined with a single hydrogen atom; in other words, a water molecule with one hydrogen atom removed.

HYPERBOLA. An extremely elongated, open-ended curve.

INERTIA. The tendency of an object at rest to remain at

rest, or of an object in motion to remain in motion, until acted upon by some external force.

INFRARED. Heat radiation.

INSOLATION. Solar radiation reaching the earth or some other celestial body.

ION. An electrically charged atom or molecule.

IONIZATION. The process of ion formation. *See* Ion.

KILOTON. One thousand tons (usually metric tons); alternatively, a release of explosive energy equivalent to the explosion of one thousand tons of TNT.

KINETIC ENERGY. The energy carried by an object in motion.

KRYPTON. A hypothetical planet from which some long-period comets are believed by some astronomers to have arisen.

MARIA. The lunar "seas," actually vast lava plains.

MASS. The amount of matter present in an object.

MESOZOIC. The intermediate age of life on earth; the so-called age of dinosaurs.

METEOROIDS. Rocks traveling through space prior to falling into the earth's atmosphere.

METEORITES. Meteoroids that survive their descent through the atmosphere and reach the ground intact.

METEORS. Luminous trails of gas left behind by falling meteoroids.

NEBULA. A cloud of interstellar gas and dust.

NITROUS OXIDE. A psychoactive chemical and anesthetic commonly known as laughing gas.

NOVA. An exploding star.

NUCLEIC ACIDS. In biology, the elementary compounds from which deoxyribonucleic acid (DNA) and other chemicals of life are made.

NUCLEUS. The solid body of ice and dust from which a comet expels dust and gas.

OCCAM'S RAZOR. The principle of logic according to which the simplest explanation of an observed phenomenon is most likely to be correct.

OORT CLOUD. A postulated swarm of billions of comets at the edge of our solar system and probably other solar systems as well; the source for comets that pass close to the sun.

OUTGASSING. The release of gases from the interior of a planet or comet nucleus.

P-WAVES. Earthquake waves that travel deep through the earth's interior.

PALEOZOIC. The earliest age of life (mostly marine life) on earth.

PARABOLA. An open-ended curve midway in configuration between a hyperbola and an ellipse. *See* Ellipse, Hyperbola.

PARALLAX. The technique used to determine the distance of a celestial object by noting the shift in its apparent position when observed from two widely separated locations.

PERIHELION. The point in a planet's or comet's orbit when it is closest to the sun.

PHOTOSPHERE. The visible surface of the sun.

PLANETESIMALS. Small celestial bodies—miniplanets—that represent either the debris of a preexisting, shattered planet or potential components of a planet that never came together.

PLANETOIDS. Celestial objects larger than a boulder but smaller than a planet.

PLASMA. A gas in which individual atoms have been stripped of their electrons.

PLEIADES. An association of seven young, hot stars in the constellation Taurus.

PRECESSION. The ''swiveling'' motion of a planet's or comet's orbit around the sun.

PROTON. A positively charged subatomic particle.

RAGNAROK. The Norse legend of the destruction of the world.

RETROGRADE MOTION. The apparent and temporary backward movement of a planet in its orbit.

S-WAVES. Earthquake waves that travel along the surface of the earth.

SECTOR BOUNDARY. The boundary between two opposing magnetic fields in space.

SHATTER CONES. Conical structures produced in rock by the passage of intense pressure waves following a planetoid impact.

SILICATE. A mineral containing silicon dioxide.

SOLAR WIND. The sun's output of particle radiation.

SPECIFIC GRAVITY. The weight of a substance relative to an equal volume of water.

SPECTROGRAM. The visual record of a spectroscope. *See* Spectroscope.

SPECTROSCOPE. An instrument that uses a prism to break down light from an observed object into its component spectrum. *See* Spectrum.

SPECTRUM. The range of wavelengths of light emitted or reflected by an object.

T TAURI STAR. A star that emits more light and heat at some times than at others.

TAIL. The stress of gas and dust extending from a comet's head in a direction away from the sun.

THRUST. Impetus for motion.

TILL. Unconsolidated sediment laid down by receding glaciers.

TSUNAMI. A seismic sea wave; often but inaccurately called a tidal wave.

ULTRAVIOLET. A band of shortwave solar radiation just beyond the violet end of the visible spectrum of light.

UNIFORMITARIANISM. The school of scientific thought in which most change in nature is supposed to be the work of steady, gradual processes operating over long periods.

VELOCITY. Distance traveled over time.

VOLATILES. The easily vaporized components of a comet nucleus.

VORTICES. Hypothetical whirlpool-like currents that

were once thought to govern the orbital motions of comets and planets.

WEIGHT. A measure of gravitational influence on mass; in simple terms, mass plus gravity.

ZODIAC. The band of constellations occupied by the planets.

ZODIACAL LIGHT. A faint glow in the night sky produced by sunlight reflected from dust left behind in space by passing comets.

Who's Who in Comet Science

ARISTOTLE (384–322 B.C.). Greek naturalist; conceived of the universe as a set of crystalline spheres centered around the earth and bearing the sun, moon, planets, and stars; imagined that comets were fiery storms in the upper atmosphere.

BARNARD, Edward (1857–1923). American astronomer; best known for his work in stellar astronomy but also a highly successful comet hunter.

BESSEL, Friedrich (1784–1846). German astronomer; observed Biela's Comet and formulated a theory to explain the behavior of comet tails.

BIELA, Wilhelm (1782–1856). Austrian military officer and amateur astronomer; helped to confirm the periodic nature of a well-known comet named in his honor.

BORELLI, Giovanni (1608–1679). Italian astronomer; helped establish correct model of cometary orbits.

BRAHE, Tycho (1546–1601). Danish astronomer and astrologer; observed comets and demonstrated that they were not atmospheric phenomena, as Aristotle had believed, but instead were residents of outer space.

CASSINI, Giovanni (1625–1712). Italian-born astronomer and comet watcher; famed for discovering the gap be-

tween Saturn's ring systems and for helping Edmond
Halley with his research on comet orbits.

CLARAMONTIUS (1565–1652). Italian astronomer (real
name Chiarimonti); published a specious attack on the work
of Tycho Brahe and used misleading data to impugn Tycho's
research on comets.

COPERNICUS, Nicolaus (1472–1543). Polish monk and
astronomer; devised the sun-centered model of the solar system that eventually replaced Aristotle's earth-centered
model.

DESCARTES, René (1596–1650). French philosopher and
mathematician; originated the "vortex theory" of the motion of celestial objects, later disproven by Halley and Isaac
Newton.

DONNELLY, Ignatius (1831–1901). American politician
and novelist; wrote a melodramatic and scientifically
worthless treatise about the effects of a giant comet passing
close to the earth.

ENCKE, Johann (1791–1865). German astronomer; discovered the periodic nature of a well-known comet that was
named in his honor.

FLAMMARION, Camille (1842–1925). French author;
wrote a lurid novel, similar to Donnelly's, in which a
gigantic comet collides with the earth and destroys it.

GALILEO (1564–1642). Italian astronomer; disputed Tycho's opinion that comets were celestial objects, saying instead that comets were merely optical illusions.

GAUSS, Karl Friedrich (1777–1855). German mathematician and inventor; devised a simple mathematical system
for working out the orbits of comets on the basis of only a
few observations.

GRASSI, Oratio (1583–1654). Italian astronomer and
clergyman; disagreed with Galileo about the nature and
motion of comets and the overall scheme of the solar system.

HALLEY, Edmond (1656–1742). English astronomer and
mathematician; collaborated with Isaac Newton in comet
studies and in the writing of Newton's mathematical works.

HEVELIUS, Johannes (1611–1687). Dutch astronomer; believed that comets were cast out of the planet Jupiter.

HOYLE, Fred (1915—). British astrophysicist; famous for originating the steady-state model of the universe and for suggesting that disease-causing microorganisms may have evolved on comets and drifted down to earth occasionally, causing epidemics.

KEPLER, Johann (1571–1630). German mystic and astronomer, Tycho's colleague; formulated the laws of planetary motion, refuted the charges of Claramontius and Galileo, and devised a model of comets' structure that is very similar to the one widely accepted today.

KULIK, Leonid (1891–1942). Soviet mineralogist and museum curator; investigated the Siberian disaster and tried to explain it as a meteorite impact.

LAPLACE, Pierre Simon de (1749–1827). French mathematician and physicist; invented an elegant but cumbersome method of calculating comet orbits.

LYTTLETON, R. A. (20th century). British mathematician; visualized comets as swarms of unconsolidated particles that released gas on close approaches to the Sun.

MESSIER, Charles (1730–1817). French comet watcher; discovered numerous comets and had one named after him.

NEWTON, Isaac (1642–1727). English mathematician and physicist; formulated a model of comets very similar to the one widely accepted today, and used the orbital motion of comets to illustrate his discoveries in mathematics as explained in his *Principia Mathematica.*

OLBERS, Wilhelm (1758–1840). German amateur astronomer; invented a simple method for calculating comet orbits.

OORT, JAN (1900—). Dutch astronomer; conceived the model of the "Oort cloud," a swarm of billions of comets circling the sun at a great distance.

ÖPIK, Ernst (1893—). Estonian astronomer; supported Oort's view of the origins of comets in a far-flung circumsolar swarm.

ORIGEN (c. 185–c. 254). Early Christian philosopher;

was among the first to propose that comets might be precursors of beneficial events as well as evil happenings.

OVENDEN, M. W. (1926–). Canadian astronomer; has proposed that some long-term comets were formed in the breakup of a planet that used to exist between the orbits of Mars and Jupiter.

PLINY THE ELDER (23–79). Naturalist and chronicler; devised an early classification system for comets.

PONS, Jean-Louis (early 19th century). French comet hunter; discovered several dozen comets in his career.

PTOLEMY (second century A.D.). Egyptian astrologer; codified astrological lore and contributed to a superstitious view of comets.

SCHIAPARELLI, Giovanni (1835–1910). Italian astronomer; established a connection between comet orbits and meteor showers.

VELIKOVSKY, Immanuel (1895–1979). Russian-born American author and pseudoscientist; postulated that a giant comet had passed close to the earth in ancient times and caused the disasters and miracles described in the book of Exodus.

VERNE, Jules (1828–1905). French science-fiction writer; used comets frequently in his novels.

WHIPPLE, Fred (1906–). American astronomer; corrected Lyttleton's erroneous model of comets by supposing that comets consist of individual, cohesive balls of ice and dust rather than the particle swarms that Lyttleton imagined.

WHISTON, William (1667–1752). English mathematician and colleague of Newton and Halley; wrote a curious biblical commentary in which he invoked comets to explain Old Testament stories.

WICKRAMASINGHE, Chandra (1939–). Sri Lankan astrophysicist; coauthor, with Fred Hoyle, of the "disease from space" hypothesis.

Recommended Readings

The literature on comets is huge, and the following list includes only a few of the books and articles about comets that non-scientists may enjoy.

Brandt, John, ed. *Comets: Readings from Scientific American.* San Francisco: W. H. Freeman and Company, 1981. This inexpensive and attractively illustrated book covers aspects of comet lore and science ranging from Giotto's portrait of Halley's Comet to the physics of comet tails and the solar wind.

———, and Chapman, Robert. *Introduction to Comets.* Cambridge: Cambridge University Press, 1981. One of the most impressive books about comets for the non-astronomer, *Introduction to Comets* provides an absorbing history of the evolution of cometary physics and is strongly recommended.

Brown, Peter. *Comets, Meteorites, and Man.* New York: Taplinger, 1976. Brown achieves the remarkable feat of conveying difficult material without boring or talking down to the reader.

Calder, Nigel. *The Comet Is Coming: The Feverish Legacy of Mr. Halley.* New York: Viking, 1981. Calder's short and lively book focuses on Halley's Comet but also presents much interesting background material on comets as a phenomenon.

Dyson, Freeman. *Disturbing the Universe*. New York: Harper and Row, 1979. Freeman Dyson's memoir of his life and career in science contains his intriguing speculations for setting up human colonies on comets.

Emerson, Edwin. *Comet Lore*. New York: The Schilling Press, 1910. This rare but delightful little pamphlet is still to be found in some libraries and recounts how comets have scared humankind from Babylonian times to the present. Emerson includes an especially hilarious account of public reaction to the visit of Halley's Comet in 1910.

Gardner, Martin. *Fads and Fallacies in the Name of Science*. New York: Dover, 1952. Gardner's classic study of pseudoscience deals with Immanuel Velikovsky's *Worlds in Collision*, Ignatius Donnelly's bizarre *Ragnarok*, and the weird comet-based theology of William Whiston.

Richardson, Robert. *Getting Acquainted with Comets*. New York: McGraw-Hill, 1967. Richardson's brief but colorful history of comet studies is an excellent, non-technical introduction to comet science.

Sagan, Carl. *Cosmos*. New York: Random House, 1980. Though Sagan's lavishly illustrated book touches only briefly on the subject of comets, it is worth reading for its account of the fall of the Aristotelian cosmology and the rise of the Copernican system.

——. *Broca's Brain*. New York: Random House, 1979. Sagan devotes a chapter to Velikovsky's hypothetical comet.

Seargent, David. *Comets: Vagabonds of Space*. New York: Doubleday, 1982. More technical than Calder's book, Seargent's study of comets contains a useful set of appendices including, among other things, a list of suggested observations for amateur comet watchers to try.

Tucker, William. "Apollos and Thunder Stones." *Star and Sky*, September 1980, pp. 36–43. An entertaining introduction to Apollo objects and the effects of their occasional impacts here on earth.

Velikovsky, Immanuel. *Worlds in Collision*. New York: Doubleday, 1950. Velikovsky's book is widely available in paperback editions. Most of his conclusions have been discredited, but some of the legends mentioned in *Worlds in Collision* are worth reading in light of what we know about the projected effects of planetoid impacts on the earth.

Index

Abraham, 12
Aeschylus, 44
Alaric, 16
Amor, 138
Anaxagoras, 45
Antipater, 13
Antitycho, 56, 60
Apollo, 138
Apollo objects, 138–41,
    147–54, 167, 168, 170,
    172, 180–86
Apollonius of Myndus, 45
Arend-Roland, Comet, 33
Aristobolus, 13
Aristotle, 45–49, 50, 54, 57
Arrhenius, Svante, 95
Assayer, The, 63–64
Asteroids, 109, 121–22
Astrology, 10, 79, 53
Astronomical Unit, 60
Atlantis, 22
Attila, 16
Aztecs, 5–8

Babinet, Jacques, 19, 102
Babylon, 10, 44
Bacon, Roger, 2
Balaam, 12
Balbillus, 14
Barnard, Edward E., 83
Bayeux Tapestry, 15
Belgrade, 18
Bell, Eric Temple, 78
Bennett's Comet, 33
Bermuda, 27
Bessel, Friedrich Wilhelm,
    33–34, 94–95, 98
Bethlehem, Star of, 9–13,
    16
Biela's Comet, 20, 33–36
Birkeland, Olaf, 97
Boniface VIII, 17
Borelli, Giovanni, 66, 72, 79
Brahe, Tycho, 51–61, 62, 64,
    77
Brent Crater, 146
Bulfinch, Thomas, 181

Cassini, Giovanni, 67, 68
Cassiopeia, 54
Caesar, Augustus, 9
Caesar, Julius, 8–9, 48
Calixtus III, 18
Canyon Diablo, 145
Cerberus, 140
Challis, James, 35
Chapman, Sydney, 97
Charlemagne, 17
Christ, 9–13, 18, 39
Chubb, Frederick, 143–45
Chubb Crater, 144–46, 157
Claramontius, 56, 60–61
Collegium Romanum, 56
Comets, influence on human history, 1–3; mentioned in English poetry, 3; inspiration to Protestant preachers, 3–4; caused terror in Catholic Europe, 4; seen as omens of future events, 5–20; involved in Spanish conquest of Aztec empire, 5–8; associated with deaths of Roman rulers, 8–9, 13–14; possible explanation of Star of Bethlehem, 9–13; inspired Massacre of Innocents, 13; played prominent role in story of Norman Conquest, 15; associated with various calamities, 15–19; comet scares in modern times, 19–21, 23–31; alarmist literature, 22–26; how comets are designated, 32–33; some famous comets, 33–43; early views on nature of comets, 44–50; *Cometographia*, 65–66; Tycho's observations of comets, 54–57; Kepler's studies of comets, 58–61; Galileo's comet studies, 61–64; Kepler's dispute with Galileo, 64; Kepler's erroneous model of comet orbits, 64–66; Hevelius's comet studies, 65–66; Borelli's comet orbits, 66; Halley's introduction to comets, 67–68; comets and Descartes' "vortex" model, 68–69; Newton's studies of comets, 72–78; periodicity discovered by Halley, 79–82; difficulties in calculating comets' orbits, 84–86; peculiarities of Encke's Comet, 87–89; improved equipment for observing comets, 89–91; chemistry of comets revealed by spectroscopy, 91; possible link between meteoroids and comets, 92; different tail configurations, 93–94; early explanations of tail dynamics, 92–96; solar wind invoked to explain tails' behavior, 96–99; structure of comets,

99–108; "flying sandbank" model, 100–04; Whipple "dirty snowball" model, 104–08; postulated origin of comets in "Oort cloud," 108–11; Krypton model of comet origins, 111–13; possible appearance of nucleus, 114–15; interaction of comet orbits with Jupiter's gravitation, 118–20; coma and tail of comet, 122–23; possible cometary origin of Martian and Terrestrial atmospheres, 124–25; breakup of comet on close approach to sun, 127; explosion of comet fragment over Siberia, 130–38; connection established between comet nuclei and Apollo objects, 138–42; large impact craters produced by Apollo objects, 143–48; energy budget of projected impact, 149–51; effects of impact, 155–64; Apollo object impact as possible cause of dinosaurs' extinction, 164–70; mythology and comet impacts, 171–86; comets as possible sources of infectious microbes, 187–93; collection of comet dust for analysis, 193–97; colonization of comets, 197–98; techniques for observing comets, 199–208

Conjunction, 10
Constantine, 16
Constantinpole, 16, 18
Copernicus, Nicolaus, 51, 54, 56–57, 58, 62
Cortés, Hernando, 5–8

David, King, 38–39
Daedalus, 140
Deep Bay Crater, 146
Democritus, 45
Descartes, René, 68–69
Donati's Comet, 36
Donnelly, Ignatius, 22–23, 174, 181

Eginard, 17
Einstein, Albert, 45, 86
Emerson, Edwin, 25, 31
Encke, Johann, 87–89
Encke's Comet, 87–89, 103, 108, 111, 141
End of the World, The (La Fin du Monde), 23–26
Euphorus of Cyme, 45

Flammarion, Camille 23–26
Fulgentius, 10

Galileo, 61–64, 69, 73
Gauss, Karl Friedrich, 86, 87
Giotto, 11
Gomorrah, 15
Götterdämmerung, Die, 181

Grassi, Oratio, 56, 62, 63–64
Gregoras, Nicephoras, 5, 78

Hagecius, 56
Halley, Edmond, 66, 67–82, 84, 172
Halley's Comet, 11, 14–15, 19, 24–25, 26–31, 32, 60, 78–82, 85, 103, 114, 197, 198, 200–13
Hamilcar, 81
Hannibal, 16, 81
Harold II, 14, 78
Harper's Weekly, 19
Hector Servadac, 21
Herschel, John, 2, 98
Hevelius, 66, 67, 74, 79, 118
Hippocrates, 44
Hooke, Robert, 69
Hubble, Edwin, 36
Hudson Bay, 146
Humason, Comet, 36–37
Humason, Milton, 36–37
Hutton, James, 165

Iceland, 168–69
Ikeya, Comet, 37
Ikeya, Kaoru, 37–38
Ikeya-Seki, Comet, 37–38
Isaiah, 17, 182–84

John, Apostle, 185–86
John of Damascus, 11–12
Josephus, 16, 98
Jubmel, 180
Jupiter, 79–80, 86, 108, 110, 118–20, 121, 139, 141

Kazantsev, Alexander, 135–36
Kepler, Johannes, 51, 58–61, 64–66, 69, 79, 84–85, 86, 99, 103
Kohoutek, Comet, 39–40, 111
Kohoutek, Lubos, 39–40
Kulik, Leonid, 131–36
Krypton, 111–13

Laplace, Pierre Simon de, 85–86, 102
Limoges, Montaigne de, 82
Lucifer, 183–84
Luke, 184
Luther, Martin, 4
Lyell, Charles, 165
Lyttleton, R.A., 100–01, 104

Macrinus, 81
Magellanic Clouds, 33
Manicouagan, Lake, 147
Mars, 86, 108, 123–24, 125, 141
Marvell, Andrew, 3
Matthew, 10, 16, 184
Mather, Increase, 4, 14
Mechain, Pierre, 82, 87
Meen, V.B., 144–45
Melanchthon, Philip, 4
Mercury, 127, 141
Messier, Charles, 82–83
Milton, John, 3
"Miners' Comet," 41
Minnesota, 22
Mitchell, B.A., 29
Mitchell-Jones-Gerber, Comet, 40

Mithridates, 16
Moctezuma II, 5–8
Mohammed, 13
Montes Appeninus, 157
Morstadt, Joseph, 34
Mrkos, Comet, 40

Napoleon I, 82, 85
Nature, 20–21
Neptune, 118
Nero, 13–14
New Theory of the Earth, A,
    172–74
Newton, Isaac, 70–81, 84,
    87–89, 99, 104, 174
Nova, 9

"Occam's Razor," 51
Olbers, Wilhelm, 86
Olson, Roberta, 12
Oort, Jan, 108–11, 113
Öpik, Ernst, 110
Origen, 12
Osecola, 19

Parallax, 55–56
Paré, Ambroise, 1, 89
Peloponnesian War, 16
Peter, 18, 184–85
Pleiades, 16
Pliny the Elder, 48–49
Pluto, 118
Pocahontas, 82
Pompeii, 16, 48
Pons, Jean-Louis, 82
Posidonius, 48
Powhatan, 82
Prescott, William Hickling, 6

Principia, 70–72, 74, 78
Proctor, Mary, 27
Ptolemy, 49, 67, 92

Quetzalcoatl, 5–8, 140

Ragnarok (legend), 181–82
Ragnarok (novel), 22–23
Ra-Shalom, 139–40
Regiomontanus, 78
Reinmuth, Karl, 138
Roman Empire, 8–9, 13–14

St. Helena, 67
St. Lawrence, Gulf of, 146
Santini, Giovanni, 35
Saturn, 53, 68, 79–81, 111,
    118
Schiaparelli, Giovanni,
    91–92, 99–100
Schickard, Wilhelm, 84–85
Secchi, Angelo, 36
Seki, Tsutomu, 37–38
Seki-Lines, Comet, 41
Seneca, 45, 48
SOLWIND, 77
Shakespeare, William, 3
Smith, Captain John, 82
Størmer, Carl, 97
Swift, Jonathan, 5, 78
Swift-Tuttle, Comet, 92

Tago-Sato-Kosaka, Comet,
    41
Tambora, 163
Tamerlane, 17–18
Tebbutt, John, 41
Tebbutt's Comet, 41

Tempel-Tuttle, Comet, 92
Thebes, Madame de, 25–26
Thirty Years' War, 19, 61
Twain, Mark, 26, 29
Twilight of the Gods, 181
Tycho. See Brahe, Tycho

Uranus, 118
Urban IV, 17

Velikovsky, Immanuel,
   174–76, 177–79, 180
Venerable Bede, 3
Verne, Jules, 21–22

Vespasian, 14
Vesuvius, 16, 49
Volsupà, 181
Vredevoort Ring, 147

Wagner, Richard, 181
West, Comet, 42–43
Whipple, Fred, 104–05, 108
Whiston, William, 172–74
William the Conqueror, 15
Wren, Christopher, 69

Zodiac, 46
Zodiacal light, 6, 127

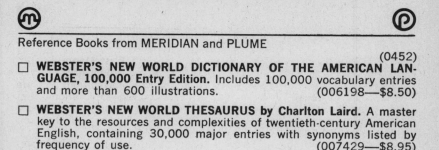

Reference Books from MERIDIAN and PLUME

☐ **WEBSTER'S NEW WORLD DICTIONARY OF THE AMERICAN LANGUAGE, 100,000 Entry Edition.** Includes 100,000 vocabulary entries and more than 600 illustrations. (006198—$8.50)

☐ **WEBSTER'S NEW WORLD THESAURUS by Charlton Laird.** A master key to the resources and complexities of twentieth-century American English, containing 30,000 major entries with synonyms listed by frequency of use. (007429—$8.95)

☐ *LOS ANGELES TIMES* **STYLEBOOK: A Manual for Writers, Editors, Journalists and Students compiled by Frederick S. Holley.** Grammar, punctuation, the general meanings and subtle nuances of words, and a wide range of journalistic techniques are among the important language tools included in this stylebook created for and used by the staff of the *Los Angeles Times.* Alphabetically arranged for quick reference and ideal for fascinating browsing. (005523—$6.95)

☐ **MAGIC WRITING: A Writer's Guide to Word Processing by John Stratton with Dorothy Stratton.** If you are a writer thinking about switching to a word processor, this complete guide will tell you in words you will understand how to choose, master, and benefit from a word processor. Plus a computerese-English dictionary and glossary. (255635—$12.95)

☐ **WRITING ON THE JOB: A Handbook for Business & Government by John Schell and John Stratton.** The clear, practical reference for today's professional, this authoritative guide will show you how to write clearly, concisely, and coherently. Includes tips on memos, manuals, press releases, proposals, reports, editing and proofreading and much more. (255317—$9.95)

All prices higher in Canada.

To order, use coupon on last page.

PLUME Fiction for Your Library

(0452)

- ☐ **THE GARRICK YEAR by Margaret Drabble.** (255902—$6.95)
- ☐ **THE MILLSTONE by Margaret Drabble.** (255163—$5.95)*
- ☐ **A SMILE IN HIS LIFETIME by Joseph Hansen.** (256763—$6.95)
- ☐ **MONTGOMERY'S CHILDREN by Richard Perry.** (256747—$6.95)
- ☐ **MA RAINEY'S BLACK BOTTOM by August Wilson.** (256844—$5.95)
- ☐ **ROYAL FLASH by George MacDonald Fraser.** (256763—$6.95)*
- ☐ **FLASHMAN by George MacDonald Fraser.** (255880—$6.95)*
- ☐ **FLIGHTS by Jim Shepard.** (255929—$6.95)
- ☐ **UNDER THE VOLCANO by Malcolm Lowry.** (255953—$6.95)
- ☐ **BACK EAST by Ellen Pall.** (255910—$6.95)
- ☐ **WALTZ IN MARATHON by Charles Dickinson.** (255937—$6.95)
- ☐ **THE FAMILY OF MAX DESIR by Robert Ferro.** (255872—$6.95)

All prices higher in Canada.

*Not available in Canada.

Buy them at your local bookstore or use this convenient
coupon for ordering.

NEW AMERICAN LIBRARY
P.O. Box 999, Bergenfield, New Jersey 07621

Please send me the PLUME and MERIDIAN BOOKS I have checked above.
I am enclosing $_____(please add $1.50 to this order to cover
postage and handling). Send check or money order—no cash or C.O.D.'s.
Prices and numbers are subject to change without notice.

Name_____

Address_____

City_____State_____Zip Code_____

Allow 4-6 weeks for delivery
This offer subject to withdrawal without notice.